CAROL WILCOX

SUGAR WATER

Hawaii's Plantation Ditches

A Kolowalu Book
University of Hawai'i Press, Honolulu

Paperback edition 1997

13   12   11   10   09   08   10 9 8 7 6 5

Library of Congress Cataloging-in-Publication Data

Wilcox, Carol, 1943–
    Sugar water : Hawaii's plantation ditches / Carol Wilcox.
      p.   cm.
    "A Kolowalu Book."
    Includes bibliographical references and index.
    ISBN 0–8248–1783–4 (alk. paper)
    1. Sugarcane—Irrigation—Hawaii—History.   2. Water resources
development—Hawaii—History.   I. Title.
SB228.W55   1996
333.91'315'09969—dc20   96–23753
                         CIP
ISBN 978-0-8248-2044-2 (pbk)

Book design by Kenneth Miyamoto

www.uhpress.hawaii.edu

The Crop

No common grass,
I am the giant of my family.
My Lordly lineage
Presents itself: Size—a tree, not a sprig.
Stalk—a solid scepter.
My elongated leaf, bowing
With slight condescension—a blade.
Flower—a frivolity. The down—an armor
Of needles, for I secret my great wealth.
Monarch monocot. I am
King Cane.

I have always ruled by inspiration,
Causing people to crave
Desserts, confections, sugar plum visions,
Fancies which, once tasted, addict.
Luxuries become necessity, a pinch
Of sugar to sweeten the day's drudgery,
And dreams of the good life, or the tolerable.
Dreams of profit
From this potent fuel for man. Dreams
As tangible as the seas, forests, mountains
Of obstacles, strewing the distance
Between you and those sweet, sweet
Dreams of King Cane.

To thrive, I make demands.
Paradise will suffice. My appetite's voracious.
I require vast and rich acreage, deep soils, full
Sun, but please, mild heat, a touch
Of evening cool, no cold.
And water.
All the water
You can find, dig, direct,
Scrounge, divert, tunnel and hold.
Bring the water tribute to me, King Cane.

Raised vigorous and ripe, towering
And impervious, I defy harvest.
You cannot squat or stoop to level me.
I am not picked or plucked, but cut
Slashed, grabbed, felled
By platoons of the strong and young,
Mobilized from every corner of Earth.
Yet, having sweated tides
To cut cane, you have no sugar.
Harness technology—ox to wheel, steam to rail,
Gas to truck—and you move cane,
But you have not made sugar.

(continued)

You have to mill me. Use
Fire and machinery. Slice and press. Scald.
Boil. Strain. Extract. Still,
I only yield my crystalline jewels
To a few grains, the seeds,
Of myself. Plant a section of my stalk,
And begin again. I am sufficient
Unto myself. King Cane.

One last sea to cross.
Sell me. The rain, wind and sun, the cane borer
And the leaf hopper are more easily controlled
Than the market.
Politics are always in order: planning,
Intrigue, an arm twisted, persuasion. Times
Change, but the game remains. Despot
To democracy, slave to laborer, owner to stockholder,
Tariffs to subsidy: these are semantic
Distinctions, and you are still my subjects.
Because in Hawaii, I rule.
Suitable successor to the burnt haul
Of sandalwood and the quick killing of whales.

You may call me by my name:
Almighty, Lordship, His Royal Highness,
King Cane.

BERYL BLAICH

Contents

Contents

Acknowledgments

In 1983, I was commissioned by the Historic Sites section of the State Department of Land and Natural Resources to conduct an inventory of Hawaii's surface water development systems undertaken by sugar plantations. These were more commonly referred to as "ditches." That survey was completed in 1984 and is on file, along with a photographic record by David Franzen, at the DLNR. It serves as the foundation for this more comprehensive history of surface water development in Hawaii.

The subject of sugar ditches spans the four major islands and a broad cross-section of Hawaii's people. Throughout my research, nearly everyone I spoke to shared a willingness to help. This was particularly gratifying as all were aware that water can be viewed from many perspectives, and that, when all else is said, water is potentially an emotionally volatile and highly political subject. My own background, with its roots in the plantations and its maturity in the environmental movement, has helped me see the history of water development from many different perspectives. One of the most sensitive (and difficult to research) of these is from that of the Hawaiian people. There is clearly much work to be done in this area. I have tried to interpret events in the context of their times, not our own. To the degree I have succeeded, I credit the wisdom and patience of the many who have assisted me. Any failures are entirely my own.

I especially wish to thank the dedicated men and women of the sugar plantations and sugar factories who allowed me access to their files, the ditch systems, the hydroelectric plants, the pumps and fields.

My thanks to the following plantations and factors for allowing me to look at their files: on Kauai—Kikiaola Land Company and McBryde Sugar Company; on Oahu—Amfac, Alexander & Baldwin, and Castle & Cooke; on Maui—East Maui Irrigation Company, Maui Land and Pineapple Company, and Pioneer Mill Company; on Hawaii—Hamakua Sugar Company.

Sugar company records are bulky and their conservation is a difficult task. Many plantation records are housed in archives. My thanks to the following institutions who made their collections available: on Kauai—Grove Farm Homestead and Kauai Historical Society; on Oahu—the Hawaiian Sugar Planters' Association, Bishop Museum, Hawaiian Mission Children's Society, Hawaii State Archives, and the Department of Land and Natural Resources; on Hawaii—the Lyman House Memorial Museum.

David Franzen's photographs of the ditches were taken in 1983 and 1984 courtesy of the following: on Kauai—McBryde Sugar Company, Kekaha Sugar Company, Lihue Plantation, Olokele Sugar Company, and Kilauea Prawns; on Maui—East Maui Irrigation Company/HC&S, Wailuku Sugar Company, Pioneer Mill Company, Maui Land and Pineapple Company; on Hawaii—Hamakua Sugar Company/Hawaiian Irrigation Company, Castle & Cooke; on Oahu—Oahu Sugar Company/Waiahole Irrigation Company, Waialua Sugar Company, Koolau Agricultural Company, and DLNR.

Many people were generous with their knowledge and time. I wish to extend special appreciation to the following: Bill Balfour, Edward Beechert, Stephen Bowles, Dick Cameron, Mary Moragne Cooke, George Cooper, Richard Cox, Richard B. Cushnie, Bill Devick, Sallie Edmunds, Lindsey Faye, Jr., Mike Faye, Roger Ferguson, Dick Frazier, Guy Fujimura, Mike Gomes, Bert Hatton, Randall J. Hee, Jack Hewetson, Don Hibbard, Arnold Hiura, Alan Holt, John Dominis Holt, John Hoxie, Bob Hughes, Charles Huxel, Tak Inouye, Irving Jenkins, Poomai Kawananakoa, Lefty Kawazoe, Marion Kelly, Damaris Kirchhofer, Tom Kunichika, Brud Larson, Kelly Loo, Ann Marsteller, Bill Meyer, Tom Nakama, Franklin Odo, Bill Paty, John Plews, Jack Poppe, Charlie Reppun, John Reppun, Monty Richards, Barnes Riznik, Barbara Robeson, Phil Scott, Pat Shade, Earl Smith, Bill Sproat, Bill Tam, Fred Trotter, Pat Tummons, Robert Vorfeld, John Wehrheim.

Thanks for historic photographs are extended to: Kathryn H. Darling for Waiahole Ditch pictures taken by her father, Ralph Heath, Oahu Sugar Company for Waiahole Ditch pictures, Maui Land and Pineapple Company for Honolua Ditch pictures taken by David Fleming, and Bishop Museum. (Original captions associated with these historic photos are set off by quotation marks.) Unattributed photographs are from the author's collection.

That this book was published at all is due to encouragement and advice at critical junctures over the last ten years from Donald Worster, Richard Cameron, David and Juliet Lee, Gavan Daws, and Iris Wiley and her colleagues at the University of Hawai'i Press. Thanks to Beryl Blaich for her poem. Thanks to my family, and especially to Gaylord, for endless patience and support.

List of
Common Standards

adf: average daily flow.

cfs: cubic feet per second—a standard measurement for the amount of water passing a given point.

mean flow: average streamflow based on all recorded flow values at a site.

median flow: flow that is equaled or exceeded 50 percent of the time.

mgd: million gallons per day—a standard measurement for the amount of water passing a given point. In keeping with the terminology common to the period when these ditches were built, measurements in the text are given in million gallons a day, although in modern times cubic feet per second is more commonly used in technical documents.

$$mgd = cfs \times 0.646$$

1 cubic foot = 7.481 gallons

1 cfs = 448 gallons per minute = 646,358 gallons per day

1 mgd = 694 gallons per minute

List of
Abbreviations

BWS	Honolulu Board of Water Supply
DLNR	Department of Land and Natural Resources
EKW	East Kauai Water Company
EMI	East Maui Irrigation Company
HC&S	Hawaiian Commercial and Sugar Company (Maui)
HIC	Hawaiian Irrigation Company (Hawaii)
MA	Maui Agricultural Company
ML&P	Maui Land and Pineapple Company
O.S.Co.	Oahu Sugar Company
USGS	United States Geological Survey

Sugar Water

Introduction

Sugar is a thirsty crop.

To produce 1 pound of sugar takes 4000 pounds of water, 500 gallons. One ton of sugar takes 4000 tons of water, a million gallons. One million gallons of water a day is needed to irrigate 100 acres of sugarcane.

When Captain James Cook came ashore at Waimea, Kauai, in 1778, he saw Hawaiians using extensive and sophisticated irrigation systems, mainly to cultivate taro. They were also growing other crops, including sugarcane.

A century after Cook—meaning a hundred years into the period of Western contact—sugar plantations started to dominate the landscape. For seventy years after that, well into the twentieth century, sugar was the single greatest force at work in Hawaii—not only economic and political but social and environmental—and water was basic to all of this.

Those were the high decades of the industrial revolution, and the story of sugar in Hawaii is the story of that revolution reaching out to a tiny group of islands in the mid-Pacific, as far from the great industrial cities of the West as any place on earth. The impact of the industrial revolution is almost incomprehensible from today's perspective. Within a few decades the world saw remarkable adaptations of steam and electric power, development of machinery and heavy equipment, scientific and technical innovations. What this meant for the sugar industry was the coming together of revolutionary field practices, higher-yielding varieties of cane, improved transportation, technical advances in the factory, and, of course, large-scale water development and irrigation. These advances allowed, for the first time, the mass production of sugar.

Hawaii had the basic natural resources needed to grow sugar: land, sun, and water. Water was a key ingredient. It was used for fluming, for the mills, and for power production. But most of all, water was used for irrigation. While sugar held promise, it needed lots of water to irrigate it on sunny leeward land

"Kaulanakaloa Flume on old ditch looking out from #13 X-cut." The ditches were re-mote and unobtrusive in the Hawaiian landscape, so their existence, scope, and impact were not generally recognized. (Photo: D. Fleming. Courtesy ML&P.)

to become more than an empty promise. The development of such enormous quantities of water might not have occurred had a less thirsty crop, say wheat or pineapple, coffee or copra, been the economic keystone. Without large quantities of water moved over long distances, the sugar industry of Hawaii would not have happened—never could have happened.

The sugar industry was the prime force in transforming Hawaii from a traditional, insular, agrarian, and debt-ridden society into a multicultural, cos-mopolitan, and prosperous one. Yet when nineteenth-century entrepreneurs in Hawaii first started looking for exportable crops, it was by no means obvious that sugar was the answer. Farmers planted taro, rice, and potatoes, coffee, oranges, olives, and silk. While some products, most notably rice, developed a small export market, there were no big successes. In 1857 there were only five sugar plantations, all doing poorly. Then sugar fortunes started to rise—due mainly to the emerging market of the West Coast as a result of the Civil War, when the Northern states boycotted Southern sugar producers and looked

Table 1
Sugar Plantations and Mills: 1884

Plantation	Location	Agents
Pepeekeo Plantation	Hilo, Hawaii	C. Afong
Wailuku Sugar Co.	Wailuku, Maui	Brewer & Co.
East Maui Stock Co.*	Makawao, Maui	Brewer & Co.
East Maui Plantation Co.	Makawao, Maui	Brewer & Co.
Onomea Sugar Co.	Hilo, Hawaii	Brewer & Co.
Paukaa Sugar Co.	Hilo, Hawaii	Brewer & Co.
Honomu Sugar Co.	Hilo, Hawaii	Brewer & Co.
Princeville Plantation Co.	Hanalei, Kauai	Brewer & Co.
Hawaiian Agricultural Co.	Kau, Hawaii	Brewer & Co.
Kaneohe Plantation	Kaneohe, Oahu	Brewer & Co.
Halawa Sugar Co.	Kohala, Hawaii	Brewer & Co.
Hitchcock & Co. Plantation	Hilo, Hawaii	Castle & Cooke
Kohala Plantation	Kohala, Hawaii	Castle & Cooke
Waialua Plantation	Waialua, Oahu	Castle & Cooke
Haiku Plantation 1 } Haiku Plantation 2	Haiku, Maui	Castle & Cooke
Paia Plantation Co.	Paia, Maui	Castle & Cooke
J. M. Alexander*	Paia, Maui	Castle & Cooke
A. H. Smith & Co.*	Koloa, Kauai	Castle & Cooke
Union Mill Co.†	Kohala, Hawaii	Davies & Co.
Kynnersley Bros.*	Kohala, Hawaii	Davies & Co.
Niulii Plantation	Kohala, Hawaii	Davies & Co.
Beecroft Plantation* Hawi Mill† Filder & Brodie's Plantation* }	Kohala, Hawaii	Davies & Co.
Waipunalei Plantation*	Hilo, Hawaii	Davies & Co.
Aamano Plantation*	Hamakua, Hawaii	Davies & Co.
Hamakua Plantation* } Hamakua Mill Co.†	Hamakua, Hawaii	Davies & Co.
Waiakea Plantation* } Waiakea Mill†	Hilo, Hawaii	Davies & Co.
Laupahoehoe Sugar Co.	Laupahoehoe, Hawaii	Davies & Co.
Kaiwilahilahi Mill	Laupahoehoe, Hawaii	Davies & Co.
Kipahulu Mill†	Hana, Maui	Davies & Co.
Barnes & Palmer*	Wailuku, Maui	Grinbaum, & Co.

Continued

Table 1
Sugar Plantations and Mills: 1884

Plantation	Location	Agents
Bailey Brothers*	Wailuku, Maui	Grinbaum & Co.
Hana Plantation	Hana, Maui	Grinbaum & Co.
Thompson & Bro.*	Kohala, Hawaii	Grinbaum & Co.
Heeia Sugar Plantation Co.	Koolau, Oahu	Grinbaum & Co.
Soper, Wright & Co.*	Ookala, Hawaii	Hackfeld & Co.
H. M. Whitney*	Kau, Hawaii	Hackfeld & Co.
R. M. Overend	Honokaa, Hawaii	Hackfeld & Co.
Kaluahonu Co.*	Koloa, Kauai	Hackfeld & Co.
W. Y. Horner*	Lahaina, Maui	Hackfeld & Co.
Chr. L'Orange*	Hanamaulu, Kauai	Hackfeld & Co.
Hanamaulu Mill†	Hanamaulu, Kauai	Hackfeld & Co.
A. S. Wilcox*	Hanamaulu, Kauai	Hackfeld & Co.
Koloa Ranch*	Koloa, Kauai	Hackfeld & Co.
Koloa Plantation	Koloa, Kauai	Hackfeld & Co.
Grove Farm*	Nawiliwili, Kauai	Hackfeld & Co.
Kilauea Plantation	Kilauea, Kauai	Hackfeld & Co.
Lihue Plantation	Lihue, Kauai	Hackfeld & Co.
Kekaha Mill Co.*	Kekaha, Kauai	Hackfeld & Co.
Pioneer Mill	Lahaina, Maui	Hackfeld & Co.
Kipahulu Plantation*	Kipahulu, Maui	Hackfeld & Co.
Waimanalo Sugar Co.	Waimanalo, Oahu	Hackfeld & Co.
R. W. Meyer	Kalae, Molokai	Hackfeld & Co.
Kukuiau Plantation*	Hamakua, Hawaii	Hackfeld & Co.
Kekaha Plantation*	Waimea, Kauai	Hoffschlaeger & Co.
Waimea Sugar Mill†	Waimea, Kauai	Hoffschlaeger & Co.
Fr. Bindt*	Eleele, Kauai	Hoffschlaeger & Co.
Makee Plantation	Ulupalakua, Maui	Irwin & Co.
Waihee Sugar Co.	Waihee, Maui	Irwin & Co.
Hawaiian Comm. & Sugar Co.	Maui	Irwin & Co.
Makee Sugar Co.	Kealia, Kauai	Irwin & Co.
Kealia Plantation	Kealia, Kauai	Irwin & Co.
Honuapo Plantation	Kau, Hawaii	Irwin & Co.
Naalehu Plantation	Kau, Hawaii	Irwin & Co.
Hilea Sugar Co.	Kau, Hawaii	Irwin & Co.
Star Mill Co.	Kohala, Hawaii	Irwin & Co.

Continued

Table 1
Sugar Plantations and Mills: 1884

Plantation	Location	Agents
Hakalau Plantation Co.	Hilo, Hawaii	Irwin & Co.
Wainaku Plantation	Hilo, Hawaii	Irwin & Co.
Paauhau Mill†	Hamakua, Hawaii	Irwin & Co.
Paauhau Plantation*	Hamakua, Hawaii	Irwin & Co.
Moanui Plantation	Molokai	Wong Leong & Co.
Olowalu Plantation	Olowalu, Maui	Macfarlane & Co.
Ookala Plantation	Ookala, Hawaii	Macfarlane & Co.
Spencer's Plantation	Hilo, Hawaii	Macfarlane & Co.
Makaha Plantation*	Waianae	Macfarlane & Co.
Waikapu Plantation	Waikapu, Maui	Macfarlane & Co.
Reciprocity Sugar Co.	Hana, Maui	Macfarlane & Co.
Huelo Mill Co.†	Huelo, Maui	Macfarlane & Co.
Grant & Brigstock*	Kilauea, Kauai	Macfarlane & Co.
Huelo Plantation*	Hamakua, Maui	Macfarlane & Co.
Kamaloo Plantation	Molokai	J. McColgan
Honokaa Sugar Co.	Hamakua, Hawaii	Schaefer & Co.
Pacific Sugar Mill	Hamakua, Hawaii	Schaefer & Co.
Eleele Plantation	Koloa, Kauai	Schaefer & Co.
Laie Plantation	Laie, Oahu	J. T. Waterhouse
Waianae Sugar Co.	Waianae, Oahu	H. A. Widemann

*Planters only.
†Mills only.
Source: Hawaiian Almanac and Annual (1884).

abroad for new sources. By 1860 there were twelve sugar plantations, all making a profit. By 1861 there were twenty-two, by 1878 there were forty-six, and by 1884 there were a total of ninety sugar planters, plantations, and mills.

The sugar industry diverted a lot of water. On Oahu, the Waiahole Tunnel delivered an average of 30 million gallons a day (mgd) and Lake Wilson yielded another 30 mgd. On Hawaii, the Kohala and Hamakua watersheds yielded 80 mgd. On Kauai, Kekaha Sugar Company brought down an average of 50 mgd, Hawaiian Sugar Company another 65 mgd, and Lihue Plantation averaged 100 to 140 mgd. The East Maui Irrigation Company's system averaged 160 mgd—and could deliver 445 mgd. By 1920, the sugar industry was diverting in excess of 800 mgd of surface water and, in addition, pumping almost 400 mgd of groundwater. The entire city of Boston used 80 mgd in 1939.

Introduction

Table 2
Fresh Water Use (mgd): by 1990

Use	State total	Hawaii	Maui	Lanai	Molokai	Oahu	Kauai
Groundwater	556.71	92.21	99.04	2.90	3.74	313.29	45.47
Domestic	134.45	18.36	19.32	0.84	0.79	86.02	9.09
Agriculture	195.42	9.31	41.80	1.96	2.36	120.67	19.29
Industry	29.18	3.95	1.85	—	—	22.90	0.48
Thermal	95.72	57.13	26.96	—	—	0.65	10.98
Commercial	101.94	3.46	9.11	0.10	0.59	83.05	5.63
Surface water	887.03	100.48	417.55	—	7.23	37.27	324.50
Domestic	1.70	0.51	0.80	—	0.12	—	0.27
Agriculture	598.17	13.52	316.10	—	7.11	37.27	224.17
Industry	22.84	16.50	—	—	—	—	6.34
Thermal	0.05	—	0.05	—	—	—	—
Commercial	0.60	—	0.60	—	—	—	—
Hydro	263.67	69.95	100.00	—	—	—	93.72
Total	1443.74	192.69	516.59	2.90	10.97	350.56	369.97

Source: State of Hawai'i Data Book (1994), table 5.18.
Note: Waiahole Ditch water is classified as groundwater.

Much of this water was diverted to Hawaii's fertile but dry plains. Hawaii's wide-open landscape was transformed by vast areas (over 200,000 acres) of green cane, plus camps, villages, towns, roads, mills, and landscaping, all made possible by the transfer of water.

Sugar's need for water varied according to the soil and climate conditions, crop maturity, and methods of irrigation. Even in those areas which received adequate rainfall, irrigated fields generally resulted in better production. By 1920 over half the fields were irrigated. Although overhead irrigation was experimented with as early as 1897, most plantations preferred furrow irrigation to minimize erosion and maximize water coverage. Fields were prepared with furrows, which ran on grade (parallel to the contour lines of the field). The cane seed was planted in these trenches and water was released into each furrow. Drip irrigation, started in the 1970s, proved to be a major improvement, increasing efficiency by some 20 percent. Almost the entire industry switched to drip irrigation. This innovation did not generally result in a reduction of water use, however. Rather it led to expanding the irrigated areas or more efficiently using the same amount of water to increase yields.

Hawaii's streams, like its native flora and fauna, are modest in size.[1] Island streams differ in many significant ways from continental waterways—in scale, water sources, and flow characteristics. Most of the water in streams comes from the mountains. Water gets to the stream quickly and directly through run-off or less directly through groundwater. Groundwater discharge—the main source of a stream's base flow—begins in the mountains and may continue throughout the stream's relatively short and often steep descent to the ocean. Geology and precipitation largely determine the location and rate of groundwater inflow to streams.

Hawaii's streams tend to be very flashy—flow rates in the streams rise and fall rapidly, often in response to rainfall. During storms these rises can be prodigious. It is not unusual for freshet and storm flows to be hundreds or even thousands of times greater than median flows. These high rates of flow occur in a short time, generally a few hours. Given the short distance from mountains to ocean, streamflow derived directly from rainfall generally reaches the ocean within a single day or two.

Almost all water withdrawals—whether by well, tunnel, or diversion—

"HSPA Waipio Substation, 4 July 1914." Monitoring devices such as these were placed on most ditches. Water measurement was critical not only to the sugar companies who used it but to the government, which charged for its use. (Photo: L. D. Larsen. Courtesy Bishop Museum.)

directly or indirectly affect streamflow. The most obvious form is stream diversion. Most stream diversions discussed here divert 100 percent of the median flow at the point of diversion, eliminating streamflow immediately below the diversion during periods of little precipitation. Higher flows usually pass over the diversion and remain in the stream. Development tunnels, which tap groundwater at higher elevations, reduce this key contribution to streams. Pumping from aquifers may also diminish the flow. By 1920, more than 800 mgd was being regularly diverted from over a hundred of Hawaii's streams, including the majority of the most productive and reliable ones. In the context of Hawaii's natural resources and cultural patterns, the scale of this movement of water was highly significant.

Ditch-related statistics should be approached with caution. Ditch statistics are generally those reported in contemporaneous accounts. But changes do occur, and the ditches of Hawaii have a life of their own. They expand, contract, enlarge, straighten, move, and change their names. Modest dirt ditches become cement-lined canals. Sometimes ditch length is measured from the intake to the first field or reservoir, sometimes down to the very last field, and sometimes to points long since abandoned. Moreover, statistics often differ slightly depending on the source.

Water quantity statistics are especially difficult to verify. Because the monitoring program was inconsistent and, moreover, was increasingly neglected over the years, reliable data were not always available. The usual sources —the USGS, the State of Hawaii Data Book, the Hawaii Sugar Planters' Association records—have all been studied and cited. But there were other variables. Often there were large differences between a plantation's capacity to divert (or pump) and the actual amount of water taken. As ditch systems deteriorated, they lost increasing percentages of water to leakage, and so the amount of water diverted and the amount of water delivered may be quite different. Unlike pumped water, which is extracted as needed, surface water diversions work all the time, regardless of the needs in the field or factory—water is not needed during rainy periods, for example, but it continues to be diverted from the streams. Because Hawaii's plantations do not have much storage capability, surplus water is usually allowed to flow to the sea.

The average, or *mean*, streamflow is based on all recorded flow values at a site. But since these values include peak streamflows, which have a large effect on computed means, the mean does not really reflect the "usual" streamflow. A more appropriate statistic for this purpose is the *median:* the flow that is equaled or exceeded 50 percent of the time (also referred to as Q50). The flow that is equaled or exceeded 90 percent of the time (referred to as Q90), is often accepted as the baseflow component of a stream's flow. Mean flow rates are more meaningful when applied to ditches, since ditches are designed with limited capacities and their flow rates do not peak as significantly as those of streams.

Sugar Water is a history of surface water development by sugar plantations primarily between 1876 and 1920, the uses for that water, the men who planned and built the ditches, and the industry it served. In Part I, we look at early Hawaiian water development and survey the principal water disputes and landmark Supreme Court water decisions. In Part II, we look at Hawaii's main ditch systems. The rare firsthand accounts of the ditch builders, often quoted at great length, provide a window on events during these decades of rapid change. The corporate squabbles on the Hamakua coast, the technical description of tunneling in Lahaina, the contractual and financial history of the Kohala Ditch Company, the working and living conditions on several projects, the raging waters in the Waiahole Tunnel, labor raiding on West Kauai—these stories are the best surviving record of how ditches in Hawaii got built.

After a hundred years of prosperity, the sugar industry's success in Hawaii appears to be coming to a close. By 1995, only a handful of plantations remained open. Lots of water was becoming available. The future of the plantation irrigation systems has become one of the most important decisions for Hawaii's own future. One forum for debating this future is the Commission on Water Resource Management, guided by the 1987 Water Code. Today the Water Commission must entertain new standards and conservation ethics when determining water allocation. It must consider an entire new vocabulary, in fact, much of which is not clearly defined in the context of water management in Hawaii, such as traditional and customary Hawaiian rights, protection and procreation of fish and wildlife, ecological balance, scenic beauty, public recreation, beneficial instream uses, and public interest. Hawaii's government and people, therefore, are facing big questions. How they are resolved will have broad economic and social implications.

Who will be responsible for maintaining (or dismantling) the ditches, reservoirs, dams, and tunnels? Who will shoulder the liability? Who will bear the cost: users or taxpayers? How can water best be used to support continued agricultural viability in Hawaii? Should it continue to be transported out of its watershed for urban use? Will water rights grounded in ancient Hawaiian law and tradition be claimed and honored? How will water conservation efforts and "best management practices" fare? Will stream restoration ever be practicable?

Just as the beginning of the twentieth century was a time of great change, so change marks the end of that century. A look at the past is essential in fact, as we step into the future. One can admire the vision and initiative of the early

By 1920 the sugar industry was diverting over 800 million gallons of surface water a day and pumping another 400 million gallons. (Photo: D. Franzen.)

sugar planters while at the same time mourning the loss of water resources and authentic Hawaiian lifestyle. The era dominated by sugar gives way to new times, new challenges, and new opportunities. Among them is a chance to manage water resources wisely for future generations. With the contraction of the endlessly thirsty sugar industry, there is now an opportunity to consider restoring a watershed management concept to Hawaii—where water is managed within the context of the *ahupua'a*, where a modern *konohiki* thinks globally, acts locally.

PART I
SUGAR AND
WATER IN HAWAII

1. Pioneers, Politics, and Profits

Nineteenth-century Hawaii was a place of opportunity, and many came to this land to try their skill or luck. Almost the entire sugar industry was peopled by immigrants: entrepreneurs, investors, engineers, laborers, skilled workers, and craftsmen came from the Pacific Islands, China, Japan, Korea, the Philippines, Europe, and the United States. This was true of the sugar industry as a whole, and of water development too. And many of these immigrants made Hawaii their home. Hawaii itself was a *pani wai*, a dam, diverting the flow of the ditch builders onto the land.

Of all agricultural products, none is more politicized than sugar. This was certainly true in Hawaii, where a strong measure of politics turned sugar from an incidental crop into a prime agricultural commodity. It was a series of political decisions that led to the Reciprocity Treaty, to sugar becoming Hawaii's economic engine, and to large water diversion projects. Industrialization set the stage for the mass production of sugar, the natural resources were available, but the show would not have gone on in Hawaii without political support.

The Hawaiian monarchy supported the sugar industry. Indeed, for years it directed major diplomatic efforts toward reducing or removing the import taxes from sugar and other products sent to the United States from Hawaii. King Kamehameha III was personally aware of the obstacles facing sugar planters, having had his own problems at his sugar plantation in Wailuku, Maui. Under this king, land was made private property and was divided among the king, chiefs, commoners, and government, in actions called the *mahele*. In 1848, the year of the first Great Mahele, the king negotiated, unsuccessfully, with the United States for trade agreements that would protect the sugar industry. His successor, Alexander Liholiho, who reigned as King Kamehameha IV from 1854 to 1863, succeeded in negotiating a treaty with the United States to allow tax-free exchange of Hawaiian and U.S. products, a treaty that the

U.S. Senate failed to ratify. King Kamehameha V continued this effort, sending an emissary to Washington in 1867. A treaty was approved by the U.S president, his cabinet, and the Hawaiian legislature, but again it was defeated in the U.S. Senate, and again in 1869 and 1871. The cession of Pearl Harbor as part of reciprocity was discussed in 1873 during the reign of King Lunalilo, who died after reigning for only a year. He was succeeded in 1874 by King David Kalakaua, who in that same year became the first monarch of any country to visit Washington, where he petitioned President Grant and the Congress on behalf of reciprocity for Hawaii.

The Reciprocity Treaty was at last ratified by the U.S. Congress, and was signed by King Kalakaua in 1876. In addition to allowing tax-free trade for most products between Hawaii and the United States, it ceded to the United States certain rights to Pearl Harbor, rights that were later expanded. The overthrow of the Hawaiian Kingdom and establishment of the Provisional Government in 1893, and the country's subsequent annexation by the United States in 1898, ensured that these mutual benefits would continue. The Reciprocity Treaty was predicated on full government support of the fledgling sugar industry, including its efforts to develop water. Without that support, which included allowing the sugar planters to transport water out of the watershed, investors would not have been attracted to Hawaii.

Upon the adoption of the Reciprocity Treaty, prospective sugar planters began at once to invest in the development of both surface and groundwater. In 1878, with Baldwin and Alexander's successful Hamakua Ditch, and again in 1879, with James Campbell's successful artesian well, it was clear that water would be available in whatever quantities were needed, to be transported wherever needed. The water development systems went by the title of "ditches." It is a term both humble and misleading: misleading because they were not all ditches—many were mostly flumes, siphons, and tunnels—and humble because their size and scale were often quite large. And they were everywhere. Very few watersheds escaped the winding, burrowing network of ditches.

The development of Hawaii's surface water was unique in that it was done almost exclusively by the private sector. Water projects in the western United States, which was undergoing a parallel history of water development, were government-funded and controlled. The Hawaiian Kingdom, ever in debt, was unable to duplicate this effort. When Maui citizens petitioned King David Kalakaua to irrigate the dry plains of the Maui "commons," Kalakaua demurred. As the government explained in its 1878 agreement with Claus Spreckels: "The Hawaiian Government is not now ready or willing to undertake such works, and incur such expenses."[1] Although a dozen years later Kalakaua investigated the feasibility of bringing water out of Hawaii's Hama-

Opening of the Olokele Ditch, Kauai. 1904. (Photo: A. Gartley. Private collection.)

kua–Kohala watershed to irrigate the Hamakua lands, the king heeded the rec-
ommendation of engineer Jorgen Jorgensen, who advised against the project.[2]
This appears to have been the sole initiative by the Hawaiian government to
develop surface water. Henceforth it was left to the private sector.

Private plantations and water companies built virtually every surface-
water collection system on the four main islands. By 1920, most large plantations had put at least $500,000 into water development, some much more. An
early legal device used by the plantations, a device that continues to the present,
is the privately held water company. These companies were established to
secure water use and access, to deal with competing water rights, to unify management of ditch systems, and to attract investors. Baldwin and Alexander were

"Drilling artesian well at Ewa ca. 1890." In 1890 the McCandless brothers dug their first of seventy-one wells for Ewa Plantation. By 1925, the total capacity of these wells was 105 mgd. Honolulu Sugar Company and Oahu Sugar Company pumped from this same aquifer. Between these three plantations they pumped over 260 mgd through almost 200 wells and produced about 175,000 tons of sugar. At that time the city of Honolulu was using about 20 mgd. According to James McCandless, speaking in 1936, "there is no other district in the world where such a great quantity of water is drawn from artesian wells in such a small area." (Photo: P. C. Lord. Courtesy Bishop Museum.)

the first to establish a private water company, the Hamakua Ditch Company, in 1876, subsequently known as the East Maui Irrigation Company. Among the other private water companies were the East Kauai Water Company, Waiahole Water Company (later Waiahole Irrigation Company), Kohala Ditch Company, Kehena Ditch Company, and a second Hamakua Ditch Company on Hawaii (later the Hawaiian Irrigation Company).

The first license to take water from streams was granted in 1876. By statute these grants are in the form of a license with specific time and payment terms. By 1915, government had granted water licenses to the private water companies noted above and to the Waimanalo Sugar Company on Oahu and the Kekaha Sugar Company and Kilauea Sugar Company (for Moloaa) on Kauai. In modern times, water licenses are granted by the Board of Land and Natural Resources (BLNR) through either a lease or a permit.

Although Hawaii's government did not develop surface water, it played a primary role in the development of groundwater, especially for domestic purposes. The sugar industry also tapped into this water reserve. The presence of groundwater on Oahu was heralded by James Campbell in 1879 with his successful well in Honouliuli. Campbell, a Scotsman, emigrated from Ireland to America when he was thirteen and to Hawaii, in 1850, at the age of twenty-four. He first invested in the sugar business in 1860, when he and two partners established Pioneer Mill Company on Maui. In 1877, he bought 41,000 barren acres on the Ewa Plains for $95,000, land worthless for sugar because of the lack of water. Campbell contracted John Ashley, a well-borer from San Jose, California, to work for him in Honolulu for one year. In July 1879, Campbell successfully tapped artesian water at Honouliuli, on the Ewa Plains. Water was developed at a depth of 222 feet; good flow was obtained at 273 feet. The well was called Wai Aniani, "Crystal Waters," and it flowed for sixty years until sealed by the City and County of Honolulu during road construction.

By 1889 there were over a hundred wells on Oahu, the largest of which rivaled the flows of the largest wells in California. Dipping a straw into the vast underground reservoirs of fresh water was often as easy as digging a hole and installing a pump—and out would come pure fresh water at the rate it was needed. In the early days, because of fuel costs, the cost of pumped water was prohibitive. But as fuel prices declined and plantations developed their own sources of power, the cost of pumped water dropped. Groundwater made extensive sugar cultivation on the dry plains of Maui and Oahu possible and accounted, too, for the success of McBryde on Kauai. Many plantations relied on groundwater to supplement their surface supply.

Sugar lends itself to large-scale production, and in Hawaii this led to a handful of agencies, known as "sugar factors," controlling the industry. Growing sugar commercially was expensive. It required costly field machinery, mills, irrigation systems, lots of land and water, lots of labor, and lots of time. (It takes two years for sugarcane to mature.) It was labor intensive. The market was not that predictable. Small plantation size and limited water supplies frequently meant failure, but large land and water projects took major capital investment. Deep pockets were very important for long-term survival. As a consequence of these economics, the sugar industry was quickly centralized.

The number of plantations, mills, and planters consolidated from ninety in 1884 to sixty in 1900. Annual sugar production increased from 140 million pounds to 540 million pounds in those sixteen years.

By 1900, about 89 percent of all plantation production was controlled by six sugar factors that had gained power through the control and management of marketing: C. Brewer & Co., Davies & Co., Castle & Cooke, Alexander & Baldwin, Irwin & Co., and Hackfeld & Co. These agencies eventually evolved into the Big Five: Alexander & Baldwin, Castle & Cooke, Theo Davies, Amfac, and C. Brewer & Company. These factors controlled the sugar industry from field to table. They owned or controlled the land, plantations, water, power production, mills, labor, transportation, refineries. They controlled banks, insurance, marketing, and, some would argue, the local government.

The factors supported the industry in many ways, especially when it came to sharing technical advice. They made sure that the quality and quantity of the product were consistent. They worked for political protection for the entire industry. By consolidating wherever possible, they reduced many of the costs. There is no question that the sugar factors were a key to the success of Hawaii's sugar industry.

The economic rewards of sugar production were great, and this spurred dramatic growth. Sugar exports nearly doubled by the decade, from 260 million pounds in 1890, to over 500 million pounds in 1900, to over 1 billion pounds in 1910, and 2 billion pounds in 1932. The value of sugar consistently delivered sweet profit, but some years were better than others. Whereas 1.2 billion pounds was valued at $54 million in 1916, 1 billion pounds was valued at $120 million in 1920. The investment in water, therefore, paid for itself many times over. The water projects were liquid monuments to the success of the sugar industry in Hawaii.

How did Hawaii's irrigation systems compare to those in the United States? By 1920, the sugar industry had invested $11 million in the development of surface water. In 1946, an economic assessment of water development concluded that "the aggregate investments (undepreciated) in major irrigation works [surface and ground] for the service of sugarcane lands exceeded $39,000,000 in 1934—an average of $304 per acre. This is considerably higher than the state average investments in irrigation works.... The next highest average capital investment for any state in irrigation works was $99 in 1940 [in Arizona]."[3]

By 1920, the industry was diverting an average of 800 million gallons of surface water daily and pumping half that much groundwater. By this time, some 250,000 acres were planted in sugarcane—and 55 percent of the crop was irrigated. Ten years later the average yield from irrigated land was 8.39 tons

"Honolua Ditch #19 East at full ditch about 80 second feet. At each tunnel portal, the pitch is similarly protected by concrete arch." Once water is diverted into tunnels and ditches, it is not always possible to restore it to the streams and watersheds of origin. (Photo: D. Fleming. Courtesy ML&P.)

per acre; from unirrigated land, the yield was 4.92 tons per acre.[4] The benefits of irrigation were evident. Unirrigated plantations were primarily on Hawaii (Puna, Kau, Hutchinson, Honokaa, and most of the Hamakua coast plantations), but even these plantations needed water for fluming and for the mills.

Over a hundred planters tried their hand at sugar production once the Reciprocity Treaty was signed in 1876. After that initial proliferation of sugar companies, the move was toward consolidation. Many found that a centralized plantation system was a more successful model than several smaller sugar planters supplying a mill. Over time, marginal companies were winnowed out, taken over, or merged with others. In 1938, there were nine large, successful companies averaging 40,000 tons or more annually: HC&S, Oahu Sugar Company, Ewa Plantation, Lihue Plantation, Hawaiian Sugar Company, Kekaha Sugar Company, Olaa Sugar Company, Waialua Agricultural Company, and Pioneer Mill. In addition to these giants, there was a handful of successful smaller companies. These survived and prospered in the golden days of sugar because of innovation, careful business management, and further consolidation.

Life on the sugar plantations bound Hawaii's people socially by creating a common cultural experience from one end of the island chain to the other. Ethnic groups were usually housed each apart from the next, and this allowed for the continuation of cultural practices within each "camp." Yet the camps were close enough that one group could observe the other's customs, foods, and rituals, and in time each group absorbed what it liked from the others. For all the differences among them, plantation people shared much in common. Plantation families all held the same jobs, wore the same clothes, woke up at the same time, had identical workdays, went home at the *pau hana* whistle to the same houses, which were even painted the same colors. People from diverse backgrounds sent their children to the one school, bought their goods at the one company store, played on the one company softball team. Hawaii's contemporary society, noted for its homogeneity, finds its foundation in this common plantation heritage.

Although the total land in sugarcane stayed around 200,000 acres, the number of plantations continued to decrease through the years. After World War II, the numbers dropped steadily: from thirty-eight in 1941 to twenty-nine in 1949, twenty in 1953, and fifteen in 1985. Of the nine large plantations in 1938 cited earlier, Ewa Plantation and Olaa Sugar Company did not survive through the 1980s. Neither of these had an extensive surface-water supply system. While water alone did not guarantee profit and survival, it was certainly a key ingredient.

By the 1980s the acreage in sugar was decreasing, too, sliding under the 200,000 acre mark in 1983. By 1995, only five plantations survived, down from fifteen just ten years earlier. On Kauai there was Gay & Robinson (which had acquired Olokele Sugar Company), Lihue Plantation and Kekaha Sugar Company (which shared management). On Maui there was Pioneer Mill and HC&S.

In more recent times, a unique agreement regarding housing was worked out between the union (ILWU) and the agencies as plantations closed. Plantations had built and maintained houses for workers, supervisors, and managers. In the 1970s, at Kilauea Sugar Company (C. Brewer & Co.) and Kahuku Plantation (Alexander & Baldwin) these residences were sold to the employees at affordable prices and financing. The consequence of this policy, as it was eventually practiced throughout Hawaii, was to endow an entire workforce with housing that would probably not have been otherwise affordable. Rarely has a failing industry provided such broad assistance to its employees.

One by one the plantations closed their doors, leaving behind isolated and closely knit plantation communities to redefine their roles in the larger island context. Fields were abandoned. Mills sat idle. Records were lost. For the most part, however, the ditches remain as legacies of phenomenal engineering and vision.

2. Water Use and Rights

The rise of sugar was not the harbinger of change to come, but rather a momentous step along the Hawaiian nation's journey from a subsistence to a Western economy. It is a dramatic story, but also a sad one. The moment of contact with Captain Cook in 1778 launched the Hawaiians into a world that they embraced or they perished. A combination of events overwhelmed the native social structure, disrupted the traditional, communal village life, and increased dependency on foreign goods and currency long before the advent of sugar. Maintaining a subsistence economy within a Western framework seemed impossible, nor is there much indication that Hawaiians desired to do so.

In a matter of a few decades, sugar economics, riding a river of water, catapulted Hawaii into the twentieth century. Fields of green cane covered previously dry plains and hills. The transfer of water to these sunny, fertile leeward lands allowed the maximization of their agricultural productivity and their urbanization. It was indeed a semblance of paradise, and immigrants flowed to Hawaii from all over the globe. Many countries coveted this island nation for themselves. The politics shifted one way and the other. The sweet spoils went, ultimately, to the United States. Although it was a mutually satisfying marriage in many ways, as both Hawaii and the mainland prospered from the relationship, it was not bliss. The flip side of this story of prosperity and growth was the native Hawaiians' continuing loss of land and water, culture and lifestyle, and the destruction of Hawaii's natural resources.

The difference between Western and Polynesian concepts of water was fundamental. Take, for example, the languages that drove the two cultures. While in English the word "water" means "a transparent, odorless, tasteless liquid, a compound of hydrogen and oxygen," in Hawaiian the word "wai" has many meanings: water and blood and passion and life. Hawaiians were fully aware of the power and wealth bestowed on those who controlled *wai.*

"Wailua Falls, 1908." Lihue Plantation tapped into the Wailua watershed for most of its water needs, reducing the Wailua Falls to a fraction of its natural flow. Hydroelectric developers propose to harness the water that still flows over the falls. (Photo: A. Gartley. Courtesy Bishop Museum.)

After all, this word is the root for the word for wealth, *waiwai,* and law, *kanawai.*

The Hawaiian subsistence economy was based on taro production. Taro was the staple of the Hawaiian diet and at the core of its culture and religion. The work it took to grow taro, to develop and maintain the irrigation systems and terraces known as *lo'i,* was shared by the entire community. As the great King Kamehameha demonstrated by working in his own *lo'i,* no man was above growing *kalo*. The survival of the Hawaiian village depended on taro cultivation, which in turn depended on shared labor and a strong, cohesive, and enduring social structure.

Although taro can be grown in dry land, it is most productive when grown in fresh, cool, shallow water, its *hā* standing straight and tall, the *wai* flowing quietly between the stalks across one terrace and down into the next. Working together, Hawaiians built *pani wai* to divert some of the *wai* from the stream, they built *'auwai* to transport *wai* to the *lo'i*, in the *lo'i* they planted *kalo*. On the banks of the *lo'i* they planted banana, ti, and *kō*, sugarcane.

Given the importance of water to taro—and of taro to the society—it follows that the native codes regarding water and its use were well established by the time Captain Cook landed in Hawaii in 1778. The *ahupua'a* was a unit of land based on watershed boundaries that served as the basis for Hawaiian land, power, and social division. Typically a valley, the *ahupua'a* was viewed as one integral unit, extending from the mountains to the sea, which included a complete complement of natural resources and ideally allowed for self-sufficiency among its residents. There was no "ownership" of water. The king's rights to water allocation were absolute. When he conveyed portions of the *ahupua'a*, he also distributed the right to use water through the authorization of *'auwai*, or ditches. Some *'auwai* systems were quite large, irrigating terraces deep in the valleys and down on the coastal plains.

The allocation of water was overseen by agents, or *konohiki*, who were also responsible for perpetuating the health of the stream itself. This was a system of great accountability. The *konohiki* lived in the village and was intimately familiar with its customs, resources, and current physical conditions as well as each individual's effort and merit. A person's right to use water was based on tradition but could be altered according to his wise management of the resource and to the competing needs of the times. When disputes over water arose it was the *konohiki* who was responsible for their resolution.

The Hawaii Supreme Court described traditional Hawaiian practices in respect to water in its Reppun Decision. The *konohiki* "endeavored to secure equality of division and to avoid troublesome quarrels between the tenants; and when the quantity of water in the stream diminished through drought he saw to it that the quantity used by each was divided equally."[1] The court further stated:

> *Perhaps the essential feature of the ancient water system was that water was guaranteed to those natives who needed it, provided they helped in the construction of the irrigation system. Because agriculture was a matter of great importance to the Hawaiians, they were, in general, willing to contribute their efforts to the water system. The konohiki aimed to secure equal rights to all maka'ainana [commoners, people in general] and to avoid disputes.*

The system based on this "spirit of mutual dependence" was a stable one. . . . The authority for the distribution of water ultimately rested in the King, the chiefs, or their agents (konohiki). . . . Interference with existing auwais was punishable by death, and the body of the offender was used to repair whatever damage done to deter further offenses.[2]

After contact with the Western world, Hawaii moved away from a subsistence (and self-sustaining) economy. Hawaii's economy was no longer based on *kalo* production, but rather on sale of goods and services. This was a profound change, not only in the economy, but in the entire society. The population shifted away from the villages and valleys to towns and seaports. Water followed the population to town. The concept of water as a resource integrated with its watershed gradually gave way to the idea that it could be transferred away from its watershed. The notion that government could take water away from the streams and springs for urban development was recognized in 1859 in "An Act to Authorize the Minister of the Interior to Take Possession of Whatever Land and Water May be Required for the Use of the Honolulu Water Works." The argument that this transfer was for the greater public good was apparently a convincing one.

But in the process of developing a municipal water supply, the Honolulu Water Works sometimes destroyed local water sources. Testimony before a water commissioner in 1873 complained that government had taken water "for the pipes. In the old times . . . there was plenty of water and it ran everywhere all the time."[3] In 1886, some Hawaiians complained that Public Works was taking water from Nuuanu Valley without regard to their rights: "The trouble now is that people get very little water on account of interference from government Lunas. . . . When water is short they do not get any—Government takes all."[4]

An 1860 statute formally acknowledged the importance of watershed protection. This act protected all government lands at the source of all streams in the district of Kona, island of Oahu, by imposing strict fines, restitution, and loss of animal for the owner of any "horse, mule, ass, hog, goat, sheep, or cattle that trespass" on streams, sources, or reservoirs.

Hawaii moved steadily through this economic transition because it always had some commodity that it could trade. At first the Orient traded for Hawaiian sandalwood; then the whaling fleet needed crew and provisions; there was California Gold Rush market; the Westerners wanted land—and these commodities all became available. Both the markets and the resources, however, were limited, and before long they were "used up." Unless it developed a new commodity, Hawaii ran the risk of becoming a political and economic nonentity, a backwater nation. This did not fit the vision that the monarch, the resident haole, or the people had for the future of the kingdom. Just as earlier

Ditch intake, Upper Lihue Ditch. Although unprepossessing, a diversion like this could handle 50 mgd. Diversions were usually equipped with adjustable gates and rubbish screens. (Photo: D. Franzen.)

monarchs had turned to natural resources to correct the balance of trade, later ones turned to agriculture. In King Kalakaua's view, the economics of sugar was compelling.

The degree to which the government believed that the sugar industry served the greater public interest was illustrated in 1876 by "An Act to Aid the Development of the Resources of the Kingdom." This measure stated that "whereas it is Advisable that the Government should Aid in Encouraging and Developing the Agricultural Resources of the Kingdom," sweeping rights of eminent domain over land and water, rights that had been previously reserved for public purposes, could now be applied by the government for the benefit of sugar. It further allowed for the lease of proposed watercourses for a term not to exceed thirty years. There was also "An Act to Regulate the Passage of Water over the Lands of Those not Benefited Thereby." This act allowed a person or business requesting a right-of-way over another's land to petition the court and clearly favored that petitioner. It included not only water transmission ditches but ditches whose purpose was to drain wetlands. For sugar to succeed, changes in uses of water were necessary. The distinction between what was in the public interest, however, and what was in the private interest became increasingly blurred. The concept of water as a public resource began to shift. To some degree, water was becoming a private affair.

The intended result of these actions was the license given to Samuel Alexander and Henry Baldwin in 1876 to "take water from the streams" for the Hamakua Ditch on Maui. Two years later the scope of this type of license was expanded to include the taking of groundwater when Claus Spreckels received permission to include the use of all the water found by "tunneling, bulkheading, crosscutting or other means such waters as may be running, flowing or percolating below the surface of the ground in gravel, sand, between bowlders, through crevices, fissures, porous tufa or lava or through any and all kinds of subterranean channels whatsoever."[5]

The shift toward Western ways appears to have been supported by the *ali'i* (the rulers) and by a significant proportion of the people as well. Perhaps the change in water use was not perceived as a major issue in the context of those disordered times. After all, the new sugar agriculture looked somewhat like the old taro agriculture: a green plant was being grown, and it too used a lot of water. But the resemblance was deceptive, especially as it pertained to water. Traditionally, most of the water diverted out of the stream to taro *lo'i* was returned to the stream, and water always remained in the watershed.[6] With the advent of the sugar industry, this was no longer the case. The sugar ditches transported enormous quantities of water permanently out of the streams—and most often out of the watershed as well.

While many people, if not the majority, benefited from these changes, they must have had an enormous impact on many others who were dependent on water in the streams. As the Hawaii Supreme Court interpreted these events many years later, "translating a system of water rights predicated upon mutual benefit into one based on private rights" was often not a comfortable fit.[7] Naturally, there was protest. In an 1895 court case we learn: "From the testimony we gather that a goodly number of other kalo patches had been abandoned by their native owners, some through disinclination to work them and others through inability to get a sufficient quantity of water to cultivate them profitably."[8]

Honomanu, Maui. Water that had supported taro cultivation in the valleys was diverted to distant sugar fields, as here, along the East Maui mountains, where several parallel ditches captured the runoff of the numerous streams for delivery to central Maui. (Photo: D. Franzen.)

The changing times called for new ways to resolve disputes. In order to address conflicts associated with water rights and the newly established right to own land introduced by the mahele, King Kamehameha IV established Commissions of Private Ways and Water Rights in each region in 1860. These Water Commissions were the official courts of appeal. A glimpse into these troubled times is provided by Water Commissioner Daniels, who said in 1866: "There is going to be much trouble in Wailuku respecting Water as the plantations are taking all the water from the natives and I am sorry to say the natives will, if it continues, become very short of Kalo for food."[9] These commissions paralleled the *konohiki* system in several important respects, above all their local familiarity and accountability.

In 1888, Kalakaua consolidated the regional Commissions of Rights of Way and Water Rights into one commission for each area. In 1907, by which time Hawaii was a territory of the United States, the statute was changed so that the water commissioners were in fact the circuit court judges and the commissions ceased to exist. This centralization of authority differed fundamentally from the traditional water management and allocation system administered by a *konohiki*. It was difficult for a protesting farmer to expect redress from a formal, distant, and impersonal court.

The public record, however, seems to include little protest over the shift of water away from the land. The records of the commissioners would no doubt provide answers to the nature and depth of protest. But with two modest exceptions, these records have not been located. Therefore, we can only speculate why the remaining record—newspapers, court documents, oral tradition—is silent on this issue. For one thing, the decline of the Hawaiian population must have been the single overriding concern of the native people. In the 100 years after Captain Cook's arrival, the Hawaiian population decreased perhaps as much as 80 percent, leaving a native people of only 60,000 in 1876.[10] Some of the causes of this are known: introduced diseases became epidemic and lethal, the birth rate dropped dramatically, many young men joined ship crews and did not return. No group was spared, and just as the young King Kamehameha II and his Queen died in London in 1824, so at home did the *kahuna*, those teachers of dance and fishing, religion, and healing, and the *konohiki*, those in charge of overseeing the management of the land and its resources, the *ali'i*, the kings, queens, and chiefs, and the *maka'āinana*, the bulk and muscle of Hawaiian society. A degree of despair, fatalism, and chaos must have characterized those times. Large numbers of Hawaiians left their traditional homes in the rural areas. By the time of sugar's ascendancy, when the large water projects were diverting water away from the valleys and their villages, these villages did not have the population, organization, or will to protest.

Another reason may have been cultural. It was unthinkable in the native culture for the people to protest what the king had mandated. As the Supreme Court explained a hundred years later: "Neither the laws of 1839 nor of 1840 were found adequate to protect the inferior lords and tenants, for although the violators of law, of every rank, were liable to its penalty, yet it was so contrary to ancient usage, to execute the law on the powerful for the protection of the weak, that the latter often suffered."[11] Small landowners were generally powerless in the face of these large-scale water decisions. Tenant farmers were usually disregarded.

Any number of other factors may have contributed to the lack of public reaction to the shift of water. In traditional Hawaiian practice, for example, many water "rights" went with the land. After the mahele, much of the land was owned by the sugar interests or by the government, which had sugar's best interest in mind. Consequently, it was hard to argue against water being put to sugar's use on these lands. Moreover, government and upstream water users (and eventually sugar interests) often took first and then waited to see if they would be brought to account. The government and later the sugar industry threatened eminent domain, a power they rarely had to use. Furthermore, in many cases the relationship between water diversion and downstream impact was gradual and delayed. By the time the diversion caused hardship, the greatest proportion of the diversion had already happened much earlier. It was hard to establish a cause and effect relationship. And, finally, the English-language press was controlled by the same interests as the sugar companies and may not have given this issue much attention.[12]

The aggrieved party could of course turn to the Water Commissions, that is, the courts. But there was always the question of how the attorneys would get paid. Justice was not for the poor in water matters. Water disputes are not generally brought to Hawaii's courts because they are expensive and time consuming. (The Hanapepe Case lasted, from first filing through appeals, from 1955 to 1991. The Reppun Case was first filed in 1976 and in 1996 was still not finally adjudicated.) Litigants in water cases must be prepared for a serious commitment of time and money. Even attorneys approach water cases with caution, especially if the client does not have a deep pocket.

In Mark Twain's view, "In the West, whiskey's for drinkin', water's for fightin'." The fighting tradition has been upheld in Hawaii. Despite all the obstacles to taking water to court, some of the most exciting—and protracted —battles have been in court, almost all of them having to do with surface water, and almost all since Hawaii has become a state.

Water moved with the times in Hawaii, from taro to sugar, from the *kono-hiki* to the court, from the village to the city, from the windward to the lee-

Nawiliwili with rice fields, 1913. For fifty years rice was the second-largest export after sugar. Ninety percent of Hawaii's rice was grown on Kauai and Oahu. Rice fields covered almost all of Kauai's lowlands, as well as the plains from Punchbowl to Diamond Head. (Photo: L. W. Hart. Private collection.)

ward, and from the public to the private. For years the Hawaiian government and then later the territorial government shared common goals with the sugar industry. During the territorial period, both the governor and justices of the Supreme Court were appointed by the U.S. president. Consequently, from 1900 to 1959, the Hawaii Supreme Court was composed of lawyers drawn from the prominent business interests whose commercial philosophy they upheld. As George Cooper summarized in his 1978 paper on the history of water rights:

> *The Supreme Court in its approximately 50 water rights decisions prior to McBryde*
> *in 1973 has a rather perfect record of developing the law in ways conducive to sugar's*

needs. The Court declared that surplus water went with the ahupuaas and ilis kuponos [sections of land] on which the waters originated making it possible for the industry to privately control most surface water sources; the Court said a water right gave the holder the power to divert the water to wherever he chose, a power crucial to sugar because most of the fields needing irrigation are distant from the water sources; adverse possession (technically here "prescription") of water rights became possible, making the stealing of a water right legal if you get away with it long enough; and early case references to riparian rights were in time weeded out or forgotten, and in any case never allowed to mature into a full-blown riparian system. Such a system, with its requirement that no one may divert outside the watershed nor take more water than would substantially diminish the natural flow of the stream, was anathema to sugar.[13]

There were small but important rulings for riparian rights. There was a series of cases on Maui from 1902 to 1904, for example, which determined that HC&S could not deprive Wailuku Sugar of water in the lowlands. On Kauai, disputes over the water of the Hanapepe River led to a divided Hawaii Supreme Court decision in *Territory v. Gay* in 1930, which found that the upper *'ili* did not have greater rights than a lower *'ili*.[14] But the most important water cases occurred long after most water diversions were in place. And these were not between Hawaiian landowners or tenants and the sugar companies, but rather between two sugar companies in one case (the McBryde Decision) and between farmers and the Board of Water Supply in the other (the Reppun Decision).

The next great change in water use and rights occurred after World War II. During the 1940s Hawaii saw the increase of the military presence, tourism, and urban population. As Hawaii became less and less dependent on the sugar industry as the only source of income, the exclusive power it had enjoyed for decades began to wane. And with that loss of influence, it was natural that the industry's apparent absolute grip on water would be rethought.

There was again a shift in government's priorities for water and, not coincidentally, the makeup of the courts. This shift became even more pronounced after statehood, which brought significant changes in the composition of the Supreme Court. It was no longer dominated by justices with interests sympathetic to sugar. The new court shifted its emphasis to acknowledge some basic Hawaiian concepts of water law by way of two landmark cases: McBryde and Reppun.

McBryde Sugar Co. v. Robinson, also known as the Hanapepe Case,[15] brought up issues of water and the public trust. In 1973, the Supreme Court handed down what is generally accepted as Hawaii's most significant water decision in the twentieth century, known as the McBryde Decision. The con-

Transporting supplies was always a major consideration on any ditch building job. Supplies were brought by men, mules, train, and boat, as here on the Honokohau Ditch. (Photo: D. Fleming. Courtesy ML&P.)

flict started in the early 1950s between Olokele Sugar Company and McBryde Sugar Company over water in the Hanapepe River on Kauai. When Gay & Robinson built the Koula Tunnel to improve the Hanapepe Ditch, this improvement worked well but in the process diminished the flow in the Hanapepe River available to downstream user McBryde Sugar. McBryde took Olokele and Gay & Robinson (and all others with water claims) to court in order to determine how much water it, McBryde, was entitled to. The parties claiming water rights were Gay & Robinson (upstream), McBryde (downstream), the Territory of Hawaii, and about ninety other individuals.

In 1973, Supreme Court Justice Kazuhisa Abe wrote the majority McBryde Decision, which looked back to Hawaiian custom, law, and intent. It found that:

> *Right to water was not intended to be, could not be, and was not transferred to an awardee by the Great Mahele and subsequent Land Commission Award and issuance of Royal Patent.*

> *. . . The ownership of water in natural watercourses, streams and rivers remained in the people of Hawaii for their common good.*
>
> *. . . Therefore, we hold that as between the State and McBryde, and between McBryde and Gay & Robinson, the State is the owner of the water in the Koula Stream and Hanapepe River. However, the owners of land, having either or both riparian or appurtenant water rights, have the right to the use of the water, but no property in the water itself.*
>
> *The State, McBryde and Gay & Robinson have both appurtenant and riparian rights to water in connection with land within the Hanapepe Valley. However, under claim of such rights, neither McBryde nor Gay & Robinson may transport water to another watershed.*[16]

The events were later interpreted by Van Dyke et al. in this way:

> *The next time the Hawaii Supreme Court looked at this question [control of surplus water] was in 1973 in the case of* McBryde Sugar Co. v. Robinson. *During the years since 1930, the political make-up of the islands had changed, and the sugar industry's economic importance had declined: Many residents had rediscovered the native Hawaiian values of conservation, sharing, and reverence for the land. The majority justices of the 1973 Hawaii Supreme Court saw a conflict between the traditional Hawaiian values and laws, on the one hand, and the Western approach which had guided the Hawaii Supreme Court in 1904 and 1930, on the other hand. They may also have been influenced by the approaches of other states to water rights, and they may have discovered that no other state in the United States appears to permit private ownership of water in the manner that the Hawaii Supreme Court had apparently approved.*
>
> *After considerable argumentation and a rehearing, the 1973 Court stated (with two of the five justices dissenting) that the Hawaiian monarchs had not intended to give away all water rights when the lands were distributed beginning in 1846, and that the stream waters belong to the State, subject to appurtenant and riparian rights. They did not rule, however, that these waters could not be used by private parties. The court based its ruling on nineteenth century statutes and invited the Hawaii legislature to consider this question more fully.*[17]

The McBryde Decision addressed appurtenant rights:

> *It is the general law of this jurisdiction that when land allotted by the mahele was confirmed to an awardee by the land commission and/or when a royal patent was issued based on such award, such conveyance of the parcel of land carried with it the appurtenant right to water for taro growing. The burden of proving the amount of water actually being used for taro cultivation at the time of the Land Commission Award is on the person claiming appurtenant water rights.*
>
> *In determining appurtenant water rights, the trial court, sitting as Commissioner of Private Ways and Water Rights, shall determine as precisely as possible the amount of water that was actually being used for taro cultivation at the time of the*

Land Commission Awards. The extent of land under taro cultivation in earlier or later time is irrelevant.

The right to the use of water acquired as an appurtenant right may only be used in connection with the particular parcel of land to which the right is appurtenant, and any contrary indication in Hawaii case law is overruled.[18]

The Hawaii Supreme Court found that the state held the water for the benefit of the people and that water could not be transferred out of the watershed. While this case applied to the Hanapepe watershed, the implications for the entire sugar industry regarding their claims to owning the "surplus" water were apparent. The case was appealed for many years. In 1989, the Ninth Circuit Court of Appeals finally dismissed a separate federal court "takings" challenge to the Hawaii Supreme Court's decision. This decision revolutionized water rights law in Hawaii.

The second Supreme Court landmark decision was in *Reppun v. Board of Water Supply*. Some of the same principles that were brought to bear in the Hanapepe Case were revisited in Reppun, this time in relation to the removal of groundwater and the effect that withdrawal had on the rights of downstream users. The Reppun Decision not only affirmed the Hanapepe Decision, but strengthened appurtenant and riparian rights. In a sense, this case went back to the time James Campbell dug his first well in 1879, for groundwater development had continued unabated since then. Much of this work was undertaken by government, particularly the counties' boards of water supply, and much of it by the sugar industry. Because of the connection between ground and surface water, this development often affected instream flow and downstream users. Typically, however, the cause and effect relationship was difficult to determine. The Honolulu Board of Water Supply's practice was to increase its pumping incrementally over the years without notification of other watershed users. At some point other users were eventually affected. The Board's right to do this was challenged in 1976 by five farmers in Waihee Valley.

Referred to collectively as Reppun and represented by the Legal Aid Society, these farmers brought suit against the Honolulu Board of Water Supply (BWS). Starting in 1955, the BWS had drilled a series of wells and development tunnels upland of taro *loʻi* in Waihee Valley in order to increase the municipal water supply to leeward Oahu. In this valley the connection between the groundwater and the water in Waihee stream is quite direct, and the BWS pumping increasingly affected Waihee's instream flow until finally Reppun could no longer get sufficient water for their *loʻi*. Reppun argued that their riparian and appurtenant rights were abridged and, moreover, that these were superior to the Board of Water's right to take water for public consumption.

The 1982 Reppun Decision, written by Chief Justice Richardson, affirmed the 1973 McBryde Decision, and by so doing laid to rest any notion that the earlier court decision was an aberration. To the contrary:

> *In McBryde, we did not lightly infer that a judicially determined system of water rights was subject to alteration. Quite to the contrary, our decision there was premised on the firm conviction that prior courts had largely ignored the mandates of the rulers of the Kingdom and the traditions of the native Hawaiians in their zeal to convert these islands into a manageable western society.*[19]

Reppun had riparian rights that were equal to, and prior to, those of the general public that the Board of Water was serving. When discussing the issue of groundwater diversions relative to surface water rights, the opinion said:

> *We therefore hold that where surface water and ground water can be demonstrated to be physically interrelated as parts of a single system, established surface water rights may be protected against diversions that injure those rights, whether the diversion involves surface water or ground water. . . . BWS's ground water diversions are therefore properly subject to limitations insofar as such diversions directly interfere with plaintiffs' riparian rights.*[20]

The case was remanded to the lower court "for further proceedings in accordance with this opinion." As of 1996 it still had not been finally adjudicated, in part because the plaintiffs lacked the financial resources to continue. Nevertheless, the Reppun Decision set the stage for returning waters to the watersheds upon which they arise.

The water diversion projects, of course, had an enormous impact on the streams themselves. Yet the degradation of Hawaii's streams cannot be attributed solely to the diversion of water—streams started drying up long before the advent of the sugar industry. Watershed destruction was a large, possibly dominant, factor. As a result of Hawaiian forest degradation, perennial streams became intermittent and springs dried up. The degradation of native forest was noted by Western observers, attributed in large part to the impact of grazing animals, first introduced by early explorers. The government recognized the need to protect the watershed and, as noted earlier, passed legislation in 1860 trying to protect the watershed from feral animals.

The intact native Hawaiian forest provided excellent forest cover. Plant cover promoted water retention—by capturing fog and enhancing infiltration or percolation—and promoted stability by preventing erosion and protecting from wind damage and heavy rain. When the forest was destroyed, rainwater was more likely to run off the land and back into the ocean, rather than percolate into the ground where it could be stored as groundwater and, among other functions, provide the stream's base flow.

The earliest tunnels were limited by the tools available—pick, shovel, ax, sledgeham-mer. The subsequent use of black powder and dynamite made extensive tunneling pos-sible. (Private collection.)

While logging and grazing had a negative impact on the Hawaiian watershed, the invasion of alien species, from plants and insects to feral animals, were likely far greater sources of watershed destruction. Feral animals destroy organic material that acts as a sponge and a basis for revegetation. As a result of vegetative loss, mass movement of sediment occurs, burying coastal reefs. And periods of drought between rainfall as well as intensity of flooding have been greater, too, as has been demonstrated on West Maui, Molokai, and Kahoolawe.

There was serious concern regarding the Kohala watershed even prior to extensive sugar cultivation or the Kohala and Hamakua ditches. The 1901–1902 drought on Hawaii's Hamakua coast was the incentive for the Tuttle Report of 1902. This document paints a bleak picture of Kohala: "Today the land is dry and unable to support life by reason of the lack of water. Old inhabitants of Kohala and Hamakua now under cane cultivation corroborate one another in stating that not many years ago there was a very large native population in those sections, and that the streams which are now usually dry were once considered as unfailing." The Tuttle Report speculated on the role of loss of native forest as a possible cause of the drought: "Of late it is claimed that, owing to the cutting of the forests, the rainfall has been diminishing and, as a result, attention has been attracted towards the possibility of irrigating the land. It has been demonstrated by continued observation in other parts of the world, that the loss of woodland has practically no effect upon the rainfall, although it does materially affect the rate of run-off."[21]

The situation did not improve. In 1924, J. Waldron, manager of the Hawaiian Irrigation Company, visited upper Hamakua and was deeply disturbed by the rapidly vanishing forest. He reported that the ohia trees were dying, that ferns were unhealthy, and that the trees below and around Puu Alala were gone, exposing "inside" country to wind and weather. Since this was government land, he suggested that the government take measures to determine the problems and then reforest. In 1926 Waldron reported seeing "a few Ficas"—the first attempts by the government at reforestation. In 1929, noting the failing Kohala forest, he associated the decreased water supply with depleted ground cover.

Despite many efforts, the government has failed to establish a replacement vegetation that is as dense or successful at retaining moisture as the native forest. Only a few vestiges of native Hawaiian forest remain.

Hawaii's water resources and watersheds have suffered through decades of being treated as unlimited resources. In recent years, however, some fundamental changes have occurred in the way people think about water-related natural resources. Rising consciousness of the importance of healthy rivers and streams, and of watershed management, has become a national phenomenon, spurred by deteriorating water quality and failure of aquatic species throughout the United States. According to the Environmental Defense Fund:

> *Americans have begun to embrace the need to protect biodiversity (the interdependent mix of living things) in such places as tropical forests and coral reefs, but the fact is that in this country the threat of extinction looms over a far greater fraction of aquatic organisms than of their terrestrial counterparts. This lost biodiversity translates into tangible losses for humans. . . . Dams, levees, channelization projects, water depletion, and the filling of wetlands have profoundly changed or destroyed aquatic habitats. The result has been species extinction, declining fish and shellfish harvests, warnings against fish consumption, and increased tumors and lesions in freshwater organisms.*[22]

American Rivers, a conservation organization, imparts an even greater sense of urgency: "Preserving our river ecosystems is absolutely critical to our survival and our economic well-being. We believe people are coming to realize that if our decisions are not ecologically sound, they cannot be economically viable."[23]

Hawaii's streams and estuaries support a small, endemic aquatic fauna that is well adapted to the unique flow characteristics of Hawaii's streams. The Hawaiians knew these fish, shrimp, and mollusks as the *ʻoʻopu nākea, ʻoʻopu hiʻukole* or *alamoʻo, ʻoʻopu naniha, ʻoʻopu nōpili, ʻoʻopu akupa* or *okuhe, ʻōpae kalaʻole, ʻōpae ʻehaʻa, hīhīwai,* and *hapawai*. These animals are diadromous—although they spend most of their lives in the streams, part of their life cycle involves the ocean. Eggs of these species are washed out to sea where they stay for four to six months as part of the ocean plankton. The larvae, or *hinano,* return to streams, possibly attracted by the freshwater plumes that enter the ocean. Thus, native aquatic species pass through the estuary twice in order to complete their life cycle. Some of these species are further distinguished by migrating miles upstream, often up steep waterfalls, and colonizing in mountain stream sections. These native species are deterred by alterations to the stream or estuary, such as channelization, reduced water flow and quality, and interruption of instream flow. Scientists are concerned about the future of healthy, reproducing populations of these animals and increasingly emphasize the "*mauka* to *makai*" connection that must be maintained for their continued viability.

In Hawaii there is a growing awareness of the public benefit of healthy streams, along with an increased understanding of the holistic connections between the forest cover, groundwater, streamflow, and coastal waters. Limits to the state's water supply are at last acknowledged. These shifts have been translated into public policy. A 1978 Hawaii State Constitution amendment affirmed the 1973 McBryde Decision:

> *For the benefit of present and future generations, the State and its political subdivisions shall conserve and protect Hawaii's natural beauty and all natural resources, including land, water, air, minerals and energy sources, and shall promote the development and utilization of these resources in a manner consistent with their conservation and in furtherance of the self-sufficiency of the State. All public natural resources are held in trust by the State for the benefit of the people.*[24]

The Constitution also mandated a water code and a water commission:

> *The State has an obligation to protect, control, and regulate the use of Hawaii's water resources for the benefit of its people.*
>
> *The legislature shall provide for a water resources agency which, as provided by law, shall set overall water conservation, quality and use policies; define beneficial and reasonable uses; protect ground and surface water resources, watersheds and natural stream environments; establish criteria for water use priorities while assuring appurtenant rights and existing correlative and riparian uses; and establish procedures for regulating all uses of Hawaii's water resources.*[25]

The 1987 State Water Code's declaration of policy described a mandate that was far different from that of the Board of Water Supply, which had been responsible simply for finding and developing sources of water at the lowest possible cost to any and all users. In stark contrast, the new water code stated:

> *Adequate provision shall be made for the protection of traditional and customary Hawaiian rights, the protection and procreation of fish and wildlife, the maintenance of proper ecological balance and scenic beauty, and the preservation and enhancement of waters of the State for municipal uses, public recreation, public water supply, agriculture, and navigation. Such objectives are declared to be in the public interest.*[26]

The code specifically included language directing the Water Commission to "establish an instream flow program to protect, enhance, and reestablish, where practicable, beneficial instream uses of water," which included fish and wildlife habitats, ecosystems such as estuaries, wetlands, and stream vegetation, outdoor recreational activities, and aesthetic values such as waterfalls and scenic waterways. Just as water shifted along with the social and economic changes of the 1880s, it is again shifting with the changes of the 1980s. Perhaps the day is coming when "water" will mean, like the Hawaiian's "*wai*," life, wealth, law, and justice.

PART II
HAWAII'S DITCHES

3. The Ditch Builders

While some of the key ditch builders were *keiki o ka 'āina*, children born of the land, the vast majority were immigrants. Among the Hawaii-born water development pioneers were William H. Rice, George N. Wilcox, Samuel T. Alexander, and Henry P. Baldwin. From points abroad arrived entrepreneurs Claus Spreckels, Valdemar Knudsen, Hans Faye, Theophilus Davies, Benjamin Dillingham, the McCandless brothers, and James Campbell. Generally without formal training, these men nevertheless conceived and executed the development of water projects to support mass production of sugar. Their vision was translated into water projects of great scope by engineers, every one of them from some foreign shore. Among them were Hubert K. Bishop, P.W.P. Bluett, J. H. Foss, Alonzo Gartley, Charles H. Kluegel, J. B. Lippencott, J. S. Malony, Joseph H. Moragne, Hjalmar Olstad, B. C. Perry, E. P. Pierce, J. Schussler, Jim Taylor, James L. Robertson, E. L. VanDerNeillen, and W. A. Wall. Among this august company of engineers, two were particularly distinguished: Michael M. O'Shaughnessy and Jorgen Jorgensen. Between the two, they accounted for seven of the largest, most ambitious projects, all built within a decade.

Michael M. O'Shaughnessy was born in Ireland and studied at the Royal University. By the time of his arrival in Hawaii in 1899, he was considered the world's foremost irrigation engineer. O'Shaughnessy engineered the 1904 Olokele Ditch, which for the first time in Hawaii replaced open ditches with long tunnels. O'Shaughnessy's legacy in Hawaii includes the 1905 Koolau Ditch on Maui, the 1907 Kohala Ditch on Hawaii and the Upper and Lower Hamakua ditches on Hawaii. As chief engineer for San Francisco, O'Shaughnessy designed and built Golden Gate Park in San Francisco as well as the Hetch-Hetchy Dam, a project reputed to have broken John Muir's heart.

Born in Denmark in 1866 and trained in the Danish military engineer corps, Jorgen Jorgensen arrived in the United States in 1889, and came to

Table 3
Hawaii Sugar Plantations: 1920–1996

KEY:

↑ merged,
  lease transferred,
  takeover, etc.
⊣ expired

A&B: Alexander & Baldwin
AF: Amfac
BT: Bishop Trust
CB: C. Brewer

C&C: Castle & Cooke
D: Dowsett
FAS: F. A. Schaefer & Co.

THD: Theo H. Davies
W: Waldron
Wh: Waterhouse

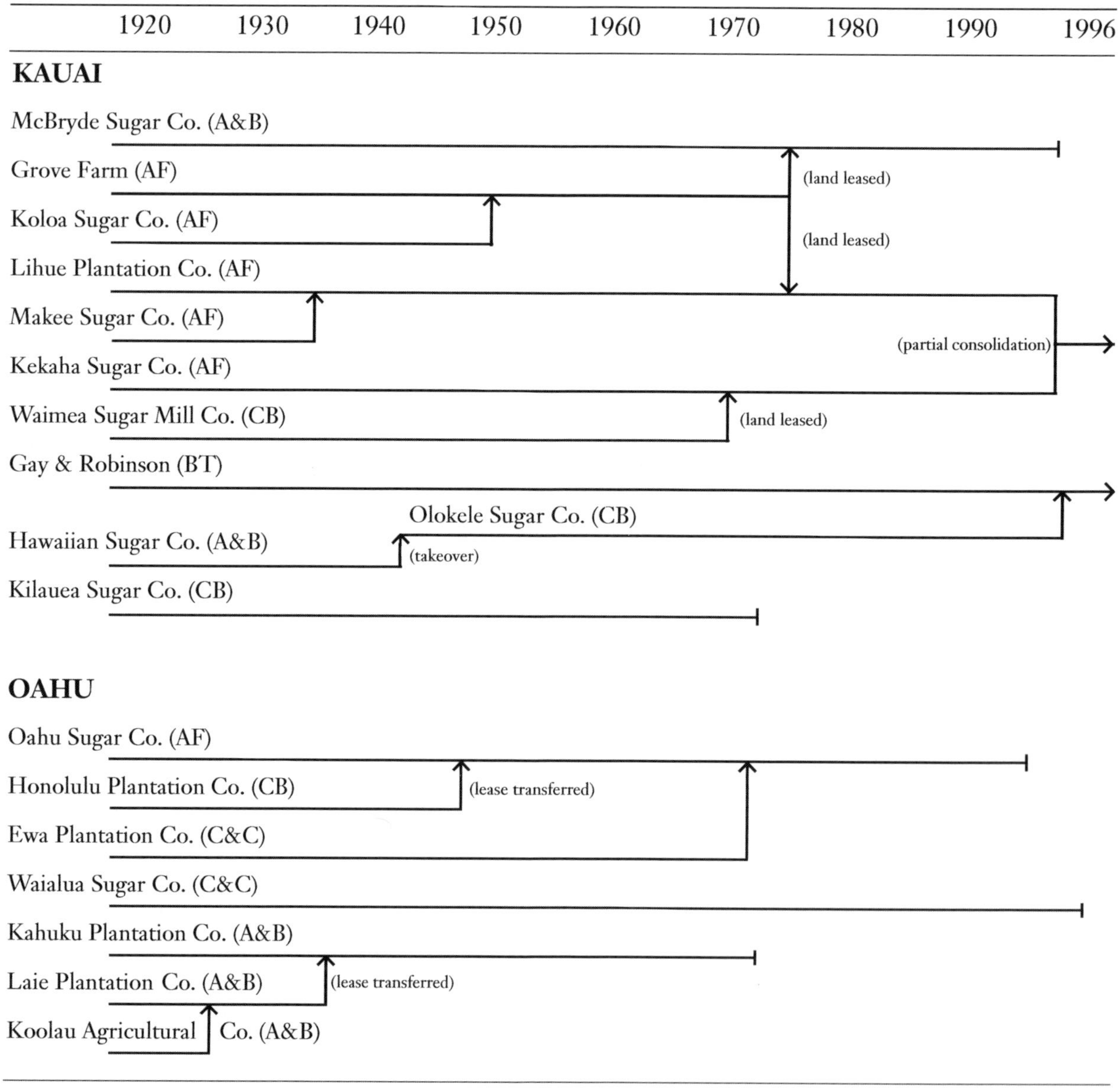

1920	1930	1940	1950	1960	1970	1980	1990	1996

KAUAI

McBryde Sugar Co. (A&B)

Grove Farm (AF)

(land leased)

Koloa Sugar Co. (AF)

(land leased)

Lihue Plantation Co. (AF)

Makee Sugar Co. (AF)

(partial consolidation)

Kekaha Sugar Co. (AF)

Waimea Sugar Mill Co. (CB)

(land leased)

Gay & Robinson (BT)

Olokele Sugar Co. (CB)

Hawaiian Sugar Co. (A&B)

(takeover)

Kilauea Sugar Co. (CB)

OAHU

Oahu Sugar Co. (AF)

Honolulu Plantation Co. (CB)

(lease transferred)

Ewa Plantation Co. (C&C)

Waialua Sugar Co. (C&C)

Kahuku Plantation Co. (A&B)

Laie Plantation Co. (A&B)

(lease transferred)

Koolau Agricultural Co. (A&B)

Table 3
Hawaii Sugar Plantations: 1920–1996

	1920	1930	1940	1950	1960	1970	1980	1990	1996

OAHU (continued)

Apokaa Sugar Co. (C&C)

Waianae Co. (D)

Waimanalo Sugar Co. (BT)

MAUI

Hawaiian Commercial & Sugar Co. (A&B)

Haiku Sugar Co.

Paia Plantation

Kailua Plantation

Kula Plantation   Maui Agricultural Co. (A&B)

Makawao Plantation

Pulehu Plantation

Kalialinui Plantation

Wailuku Sugar Co. (CB)

Pioneer Mill Co. (AF)

Olowalu Co. (AF)

Kaelaku Plantation Co. (CB)

(continued)

Table 3 (cont.)
Hawaii Sugar Plantations: 1920–1996

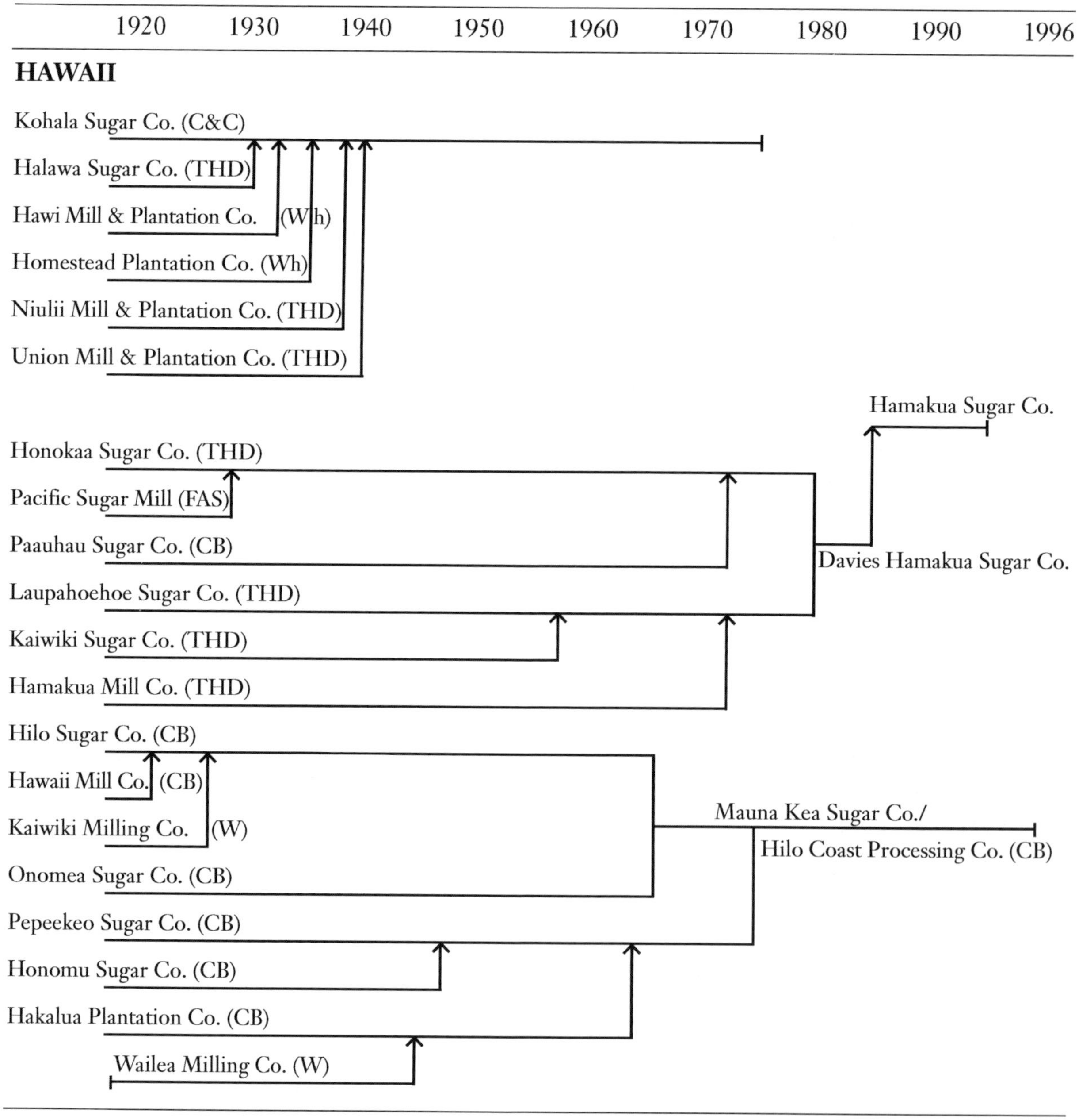

	1920	1930	1940	1950	1960	1970	1980	1990	1996

HAWAII

Kohala Sugar Co. (C&C)

Halawa Sugar Co. (THD)

Hawi Mill & Plantation Co. (Wh)

Homestead Plantation Co. (Wh)

Niulii Mill & Plantation Co. (THD)

Union Mill & Plantation Co. (THD)

Hamakua Sugar Co.

Honokaa Sugar Co. (THD)

Pacific Sugar Mill (FAS)

Paauhau Sugar Co. (CB)

Laupahoehoe Sugar Co. (THD)

Davies Hamakua Sugar Co.

Kaiwiki Sugar Co. (THD)

Hamakua Mill Co. (THD)

Hilo Sugar Co. (CB)

Hawaii Mill Co. (CB)

Kaiwiki Milling Co. (W)

Mauna Kea Sugar Co./

Onomea Sugar Co. (CB)

Hilo Coast Processing Co. (CB)

Pepeekeo Sugar Co. (CB)

Honomu Sugar Co. (CB)

Hakalua Plantation Co. (CB)

Wailea Milling Co. (W)

Table 3 (cont.)
Hawaii Sugar Plantations: 1920–1996

	1920	1930	1940	1950	1960	1970	1980	1990	1996

HAWAII (continued)

Olaa Sugar Co. (AF) ————————————————— Puna Sugar Co. ———————

Hutchinson Sugar Co. (CB) ——————————————————

Hawaiian Agricultural Co. (CB) —————————————————— Kau Sugar Co. (CB) ———————

Waiakea Mill Co. (THD) ———————————————

Kona Development Co. (Wh) ————

Ditches and flumes built along the contour of the slope were subject to storms, landslides, overgrowth, and deterioration. Most of them were replaced by tunnels, as were these on the Honokohau Ditch. (Photo: D. Fleming. Courtesy ML&P.)

Hawaii in 1898. He was engineer for McBryde Sugar Company and then Koloa Sugar Company. He supervised both the 1905 Koolau Ditch on Maui and the 1906 Kohala Ditch on Hawaii under O'Shaughnessy. He was the chief engineer for the Big Island's Upper and Lower Hamakua ditches and reservoirs from 1907 to 1911. He designed the 1915 Waiahi-Kuia Aqueduct on Kauai. He sat on the Honolulu Water Commission and was the engineer for the Hawaiian Homes Commission.

Although Hawaii is geographically remote, these entrepreneurs and engineers were not isolated from technological advances made elsewhere in the world, nor were they hesitant to use them. Not only were the engineers big thinkers, they were also quick. No project took more than two and a half years to build, including Oahu's Waiahole Tunnel—a 2.7-mile bore through the Koolau mountain range, and the 8.9-mile tunnel on the Lower Hamakua Ditch.

While it took only a handful of men to conceive, survey, and engineer these ditches, it took laborers by the thousands to push tunnels through mountainsides, flumes across valleys, and ditches to the plains. On the East Maui mountains, many hundreds of men were employed almost continuously for fifty years building the 74 miles of canals and ditches named Hamakua, Haiku, Manuel Luis, Center, Lowrie, Koolau, New Haiku, Kauhikoa, Wailoa. Labor constituted the largest single expense of all the ditches.

Finding workers in Hawaii proved challenging from the start. There was a chronic shortage of workers until the war in the 1940s, when mechanization, unionization, and other factors reduced the need for such a large labor force. The labor shortages were surely magnified during the construction of the great ditches—reference is made to 500 to 600 and 1000 men working on each of these projects for a year or more at a time. The common perception is that Hawaii's ditches were built by Chinese under such brutal working conditions that many laborers died. While this turns out to be not entirely accurate, the real story is just as compelling. In fact, the majority of the ditch building workforce was Japanese, primarily because Japanese comprised the vast majority of the general laboring population during the time of the great ditch projects.

Japan was experiencing rough times in the 1880s. Although its population was indebted and impoverished, the Japanese government forbade its citizens to work overseas in debt peonage situations. In Hawaii, however, the Masters and Servants Act of 1850 forbade debt peonage and required that labor be paid in cash—making Hawaii at the time the only place in the world with a cash wage economy. (The United States did not outlaw debt peonage until 1898.) After long resisting foreign attempts to use Japanese labor, the Japanese government changed its policy in 1885, at least relative to Hawaii. Once committed, the government proved to be a most effective advocate for its people. The

"Hand drilling." Tunnel work was often contracted for by the foot, the rate depending on the conditions. The work was often dangerous, especially when blasting was involved. While accidents were not frequent, they were always a threat and they did occur. (Photo: D. Fleming. Courtesy ML&P.)

1885 treaty between Japan and Hawaii governing this immigrant labor offered an unprecedented level of protection to the Japanese worker. The treaty required that the workers be paid in silver; the balance remitted to Japan had to be in gold. From 1885 to 1894, the Japanese government even kept the accounts in order to ensure that all the conditions were met.

The primary motivation for most of these laborers was to send money home to get their families out of debt—debt that for some had accrued over many generations. Over a million dollars in gold yen was transferred back to the prefectures in Japan, money that not only eliminated the debts but funded the establishment of a substantial middle class. In 1894 the government returned to its original policy and no longer encouraged emigration. In 1906, it ended it.

The Japanese government screened those allowed to go. The government sent twenty-five hand-picked groups between 1885 and 1894. Competition was fierce: the first group of 961 laborers was chosen from 18,000 applicants. There may be no other example in history of a group of immigrant laborers so highly motivated, skilled, and literate. About 55 percent of the 200,000 Japanese who came to Hawaii between 1886 and 1924 returned to Japan. Of those who remained, some moved on to other places but many settled in Hawaii.[1]

Starting in 1885 and throughout the major ditch building period, Japanese workers made up the majority of the ditch building labor force. Working on the ditches provided advancement opportunities for independent workers, especially those skilled in explosives, mechanics, transportation, or supervision. Over the years a skilled workforce developed and was in keen demand. Ditch building appeared to be men's work; there is no reference to women on the job. The outer islands had greater difficulty in keeping workers away from Honolulu's city lights. And despite strong efforts by the sugar factors to discourage such practices, labor raiding between plantations was not uncommon.

Once the ditches were completed, maintenance was performed by "ditchmen" stationed permanently or for extended tours in cabins along the ditch. Routine maintenance was frequently punctuated by repair of freshet and storm damage. Repairs were sometimes sizable projects. In Kohala, 1946 was an especially bad year—when 9.94 inches of rain fell in January at Honokane, washing out the Armco iron flume in the west branch of the Kohala Ditch. It took twelve days to transport the lumber through the tunnels on two-wheel buggies and build a new wood flume. There was a second storm in December, as well, that wiped out many of the telephone lines, pipelines, intakes, dams, and bridges.

Mules were used during the construction phase of many of the ditch systems. On the Kohala project: "About 250 draught and pack mules were used on this work, the native Hawaiian mule being very handy and durable for packing purposes."[2] Once the project was completed, a stable of mules was maintained. These animals, the backbone of the maintenance crew on the Kohala Ditch, were more surefooted, and stronger, than the horse and were less apt to get spooked on the narrow and treacherous trails. Since the trail had to be wide and sturdy enough to accommodate these animals, the Kohala Ditch trail was well maintained.

As labor costs rose, ditch maintenance suffered. In 1925 the Hawaiian Irrigation Company's records reflected large losses, resulting in a directive that maintenance expenses were to be "cut to the bone." The results of this move were impressive. In 1925 the Upper and Lower Hamakua ditches averaged sixty-eight laborers working twenty-three days a month, costing the plantation $3300 a month. By 1929, the same area was covered by an average of twenty-

Camp 10, Kokee. Plantations maintained cabins along many of the ditches. In some cases these housed resident ditch keepers; in others they served as shelters for maintenance crews. This system was abandoned around the 1940s due to cost, and only a few cabins remain. (Private collection.)

three workers working seventeen and a half days a month, costing the company $888 a month.

Eventually all of the plantations and irrigation companies eliminated their residential ditchmen positions. In time the cabins were abandoned and the ditchmen disappeared. Starting in the 1950s, the standard method of maintenance was by a work crew that would regularly check the ditch and trails, make necessary repairs, and stay in the cabins (if they were habitable) only as long as necessary. In speaking of the ditchmen of Waiahole Ditch, L. H. Herschler had this to say:

> Initially about 40 ditchmen and other personnel were required to operate the system. Today [1966], with better transportation, improved communications and electronic gages and gates, and with the use of herbicides for trail maintenance, the system is operated by the equivalent of nine full-time employees. This reduction in employees has come about by natural attrition. Several of our present employees are sons of fathers who formerly worked for Waiahole Water Company. The employees are non-unionized, have an outstanding esprit-de-corps and take considerable pride in their work and in the Company.[3]

4. Early Efforts

Early nineteenth-century sugar irrigation efforts did not differ remarkably in scale or technology from traditional Hawaiian practices. The 1856 Rice Ditch, for instance, was only about 10 miles long and 2.5 feet wide by 2.5 feet deep. It was built by Hawaiians who earned 25 cents a day. In order to make it watertight, "the men tramped it all over to make the bottom hard."[1] The main difference between the Hawaiian 'auwai and the early sugar ditches was that now water was not returned to the stream and in fact was transferred for the first time out of its watershed.

William Harrison Rice, founder of Lihue Plantation, pioneered modern sugar irrigation when he built the Rice Ditch on Kauai. It was opened on 16 August 1856, one of the landmarks in the history of Hawaii's sugar plantations. Starting at Hanamaulu stream, it led southward toward Lihue through a gap in Kilohana crater. W. H. Webster was the engineer and W. H. Pease the surveyor. This ditch cost about $500 a mile. The unlined and well-arched tunnels were still in use in the 1930s.

How efficient was the Rice Ditch? An inspection three days after the opening indicated no more than 2 or 3 inches of water in it. A later account, in September 1857 in *The Polynesian*, reported the Rice Ditch a success. Later events, however, suggest that it was not the success it might have been. It was too small and unnecessarily long. It was porous, too, so a lot of water was lost through seepage. It could not deliver water during drought conditions. Even so, these shortcomings in no way diminished its significance. Planters no longer had to look for the perfect conditions; now they could create them. They could bring mountain water to the hot sunny fields of thirsty cane. Suddenly there were very few places in Hawaii that were not suitable for sugar.

Two observers in particular were impressed—George Norton Wilcox, also of Kauai, and Samuel Alexander, from Maui. Within the next decade both these men would launch successful careers in sugar and would themselves become part of the history of irrigation in Hawaii.

Samuel Alexander and Henry Perrine Baldwin, both descendants of Maui missionary families, built the first large water project, the Hamakua Ditch, on Maui in 1878. Growing sugar on the dry slopes and plains of Maui was a marginal business. Within sight, however, were the tantalizing rains on the upper slopes of both the east and west Maui mountains.

Alexander and Baldwin began their business relationship on Waihee Plantation as manager and assistant. Apparently the two men were already experimenting there with irrigation, judging by a brief, intriguing reference to "a broad and deep ditch four miles long" built in 1866.[2]

Samuel Alexander was no stranger to irrigation. He was raised in Lahaina on the banks of the stone-lined Lahainaluna Ditch, used for irrigation at Lahainaluna School. When on a visit to Lihue Plantation on Kauai in 1856, he recorded his impressions of the newly completed Rice Ditch in a letter to his brother James: "Owing to the terrible drought of the past year, the Lihue Plantation has lost an entire crop. Mr. Rice has just completed an aqueduct ten miles long by means of which he can irrigate his cane. It cost about $500 a mile."

In 1876, Samuel Alexander, manager of Haiku Sugar Company, proposed to the stockholders that they undertake the building of a major ditch. They agreed. In September 1876, Alexander secured rights from the government of King Kalakaua to collect water from the slopes of Haleakala to the east of Haiku Plantation, between Honopou and Nailiilihaele streams. The license required that the project be completed in two years or all improvements would revert to the government. The Hamakua Ditch Company, forerunner of the East Maui Irrigation Company, was organized and owned by the Haiku Sugar Company, T. H. Hobron/Grove Ranch Plantation, and James and Samuel Alexander and Henry Baldwin.

In September 1876, Samuel Alexander wrote to "Dear George" Wilcox, at his Grove Farm Plantation on Kauai:

Were it possible I would beg of you to come up here to Haiku and help us engineer a new ditch, we are about starting. . . . The new water ditch I speak of is going to be the making of Haiku Plan. But it will cost from $25,000 to $30,000. I propose to bring water enough to irrigate from 2000 to 3000 acres of land. There is no question about the water supply from which I am going to draw being ample. The only question is whether the ditch I propose digging is big enough to carry what water I want. The ditch, including all the bends, will be about 25 miles long, and we propose giving it a uniform grade of 2 in. to the chain (60 ft.). It is to be 6 ft. broad on top—3½ on

Grove Farm's Lower Ditch. The earliest sugar ditches did not differ from the Hawaiian *'auwai* in technology or size. The major departures from traditional Hawaiian practices were the permanent removal of water from the stream and the transfer of water out of the watershed. (Photo: D. Franzen.)

the bottom and 3½ ft. deep. If this will not carry all the water we want we can enlarge it. We have got a great many shallow, & a few deep ravines to cross.[3]

"Now George," he continued, "I know that you are not much of a letter writer, but I want you to sit down & give me all the information on the subject of ditching you can." We do not know if George ever sat down and gave his all, but we do know that he did not go to Maui to help engineer the new ditch.

The Hamakua Ditch was built by laymen. Samuel Alexander's brother, James, did a preliminary survey which determined that the project was feasible. It is fairly certain that Baldwin surveyed its course and personally supervised its construction, and that his superintendent was a carpenter. The *Hawaiian Annual* published a description of the project in 1878.

The digging of the ditch was a work of no small magnitude. A large gang of men, sometimes numbering two hundred, was employed in the work, and the providing of food, shelter, tools, etc., was equal to the care of a regiment of soldiers on the march. As the grade of the ditch gradually carried the line of work high up into the woods, cart-roads had to be surveyed and cut from the main road to the shifting camps. All the heavy timbers for flumes, etc., were painfully dragged up hill and down, and in and out of deep gulches, severely taxing the energies and strength of man and beast, while the ever-recurring question of a satisfactory food supply created a demand for everything eatable to be obtained from the natives within ten miles, besides large supplies drawn from Honolulu and abroad.

At the head of the work many difficult ledges of rock were encountered, and blasting and tunneling were resorted to—to reach the coveted water. While work on the ditch was thus progressing, pipe makers from San Francisco were busied rivetting together the broad sheets of iron to make the huge lengths of tube fitted to cross the deep ravines. These lengths had each to be immersed in a bath of pitch and tar which coated them inside and out, preserving the iron from rust, and effectually stopping all minute leaks. The lengths thus prepared being placed in position in the bottom of the ravines, the upright lengths were fitted to each other (like lengths of stovepipe) with the greatest care, and clamped firmly to the rocky sides of the cliffs. Their perpendicular length varies from 90 feet to 450 feet; the greatest being the pipe that carries the water down into, across, and out of Maliko gulch to the Baldwin and Alexander, and Grove Ranch Plantations. At this point every one engaged on the work toiled at the risk of his life; for the sides of the ravine are almost perpendicular, and a "bed" had to be constructed down these sides.[4]

Two years earlier Baldwin had lost his arm while checking the gauge of the rollers of his sugar mill in a misfortune that seems to have in no way inhibited his riding, hunting, working, or writing. His accomplishments became part of local lore:

Tunnels were generally unlined and ranged in shape from square to round to "well arched" like this one on the Haiku Ditch at Uluwini, Maui. (Photo: D. Franzen.)

A wood-stave siphon under construction, possibly on the 1889 Hanapepe Ditch. Through the use of inverted siphons, water could indeed be made to travel uphill, provided that the grade at the exit was below that at the intake. Flumes, either built on the contour or spanning the gulch, were the other method of crossing gulches. (Private collection.)

Keeping workers provisioned was a challenge in the remote mountains. Installing a siphon on Kauai, the work gang, along with a couple of dogs and the pack mule, takes a break. (Private collection.)

Setting the siphon pipe in the Maliko gorge required the workers to go down the side of the precipice on a rope. That they refused to do. Henry Baldwin, clutching the rope with his legs and his one arm, then went down, which so shamed the men that they followed him down the rope, then and thereafter, until the job was done.[5]

The Hamakua Ditch, started in 1876, serviced the Haiku fields by July 1877. It was not lined, and for the most part traveled through an underlying stratum of red clay that made a compact and watertight bed while in other places the ground was packed to make it hold water. A contemporary report indicates "but little waste of water from soakage, far less than was anticipated, and the pipes none at all from leakage."[6] The ditch was not completed until the last days of the deadline imposed by the government lease—on 30 September 1878—by which time it had been extended to Nailiilihaele stream, intercepting the Kailua, Hoalua, Huelo, Hoolawa, and Honopou streams as well as smaller streams along the way. The costs of water projects in Hawaii were consistently underestimated, and Alexander's estimate was no exception. The length of the new ditch was only 17 (not 25) miles; its cost was $80,000 (not $30,000).

Besides the hazard of spanning Maliko gulch, Baldwin and Alexander were facing equally fearsome obstacles on the political front. There was no greater challenge than that posed by Claus Spreckels, who was to build the Haiku (Spreckels) Ditch. Claus Spreckels came to the Kingdom of Hawaii in 1876. He controlled the sugar refinery operation on the West Coast and hoped to gain control of the cane production side of the industry as well. He became friend and adviser to King Kalakaua, aligning himself with the king against the emerging sugar planters. Spreckels granted loans to the financially overextended monarch. Control of water on East Maui quickly became the focus of a dramatic struggle pitting King Kalakaua and Claus Spreckels against Sam Alexander and Henry Baldwin.

In 1878, Spreckels acquired lands on the central Maui plains to start a new sugar plantation. He bought an undivided interest in 16,000 acres of the Waikapu Commons from Henry Cornwell and leased 24,000 acres of adjacent Wailuku Commons crown lands from the government for $1000 a year. Several years later, through a process that smacked of corruption and deals, the legislature granted these lands to Spreckels in fee.

Spreckels lost no time petitioning the government for water rights to irrigate his new plantation. Kalakaua, in one of the most controversial acts of his reign, and after a late-night meeting with Spreckels and others in a hotel, sent a messenger at two in the morning dismissing his cabinet and installing a new one. This new cabinet granted Spreckels his water rights the following week. A loan from Spreckels to the king was executed that same day.

A most revealing provision of the lease gave Spreckels the right to all water not already in use at a certain date (30 September 1878)—a date that corresponded with the completion requirement date for the Hamakua Ditch. This meant that if Alexander and Baldwin's Hamakua Ditch was not finished on schedule, Spreckels could lay claim to that water and possibly the Hamakua Ditch as well. Considering the delays being encountered at Maliko gulch, this was a good possibility. Nevertheless, the Hamakua Ditch was finished in September 1878, a few days within the time limit set by the lease. Alexander and

Baldwin's ditch and water were apparently safe from Spreckels' long and daring reach.

Once it was clear that the Hamakua Ditch would be completed on time, Spreckels proceeded to build his own ditch on East Maui, which went from Honomanu stream to the Kihei boundary. This Haiku (Spreckels) Ditch was started in November 1878 and completed the following year. Spreckels used this water to irrigate his new Hawaiian Commercial Company cane lands on the central Maui plains. Jacob Adler's biography of Claus Spreckels describes the Hamakua Ditch:

> This "grand piece of engineering" involved some thirty miles of ditch, tunnels, pipes, flumes and trestles. It crossed thirty gulches, some over 2,000 feet wide and 400 feet deep. Twenty-eight tunnels, 3 by 8 feet, some of them 500 feet long, had been cut through solid rock. Twenty-one thousand feet of pipe had been used. The ditch itself was 8 feet wide by 5 feet deep, with a fall of about 3 feet to the mile. It delivered up to 60 cubic feet of water a second. All this involved an outlay of about $500,000, the largest amount spent on any irrigation project in the Hawaiian Islands up to 1880.[7]

Spreckels redefined large-scale irrigation in Hawaii. His 1879 Haiku Ditch was almost twice as long, three times as large, carried half again more water, and was six times the cost of the 1878 Hamakua Ditch. Spreckels was the first to employ a foreign engineer, Hermann Schussler, as did almost every subsequent large project in Hawaii. Adler concludes:

> The Spreckels ditch invites comparison with the earlier Hamakua ditch of Baldwin and Alexander. The Hamakua ditch was 17 miles long, delivered about 40,000,000 gallons of water a day, and cost $80,000. The Spreckels ditch was 30 miles long, delivered about 60,000,000 gallons a day, and, as already noted, cost about half a million dollars. This ditch was built makai ("seaward") of the Hamakua ditch. Both were pioneering works, but the Hamakua ditch was the first large irrigation project on the island of Maui. It is to Spreckels' credit as an entrepreneur that he promptly evaluated and grasped the importance of the Baldwin and Alexander work. He imitated its good features, profited from their experience, and avoided their mistakes.
>
> Considering the limited resources at their command, the work of Baldwin and Alexander must be judged the greater achievement. It was harder for them to raise $80,000 than it was for Spreckels to raise $500,000. They could not afford the competent engineering talent represented by Hermann Schussler.[8]

Of these two early ditches, not a great deal remains.

In 1882, Claus Spreckels reorganized his Hawaiian Commercial Company into the new Hawaiian Commercial and Sugar Company (HC&S), a California corporation. For the next seven years there was a struggle over ownership of the company between Spreckels, his sons and associates, versus the

Maui sugar planters. In 1898 Spreckels lost control of HC&S to the Maui sugar planters. In the meantime, in 1882, he built the Waihee Ditch, later known as the Spreckels Ditch, on West Maui. It started at the 435 foot elevation of the Waihee stream, had a 60-mgd capacity, and went 15 miles to Kalua, Wailuku, where it emptied into HC&S's Waiale Reservoir. This ditch, in combination with the earlier Haiku Ditch, gave Spreckels the distinction of becoming the first to irrigate his fields by water from both the East and West Maui mountains.

It was a span of only twenty years from the time a few Hawaiians built the 1856 Rice Ditch and "tramped it all over to make the bottom hard" to when 200 men labored on Baldwin and Alexander's Hamakua Ditch on Maui. Although only twenty years separated them, these two projects were vastly different in their scope and impact. Because of advances in the technology of tunneling, siphons, and flumes, ditches were no longer hostage to the terrain.

The decades leading up to the 1880s were perhaps like no other in human history. Advances in technology revolutionized people's ability worldwide to alter the environment, to produce and transport things, to communicate. And these innovations all had an impact in Hawaii, as well. The Lahaina variety of cane was brought from Tahiti in 1854. Vacuum pans were first used in 1860. In 1878, only two years after Alexander Graham Bell first spoke to Watson on the telephone, Charles Dickey strung the first telephone line on Maui between Kahului and Wailuku. In 1879, James Campbell drilled the first successful artesian well in Hawaii on Oahu; in 1881, the McCandless brothers did the same on Maui. In the 1880s, for the first time, electric lights were available, some even powered by hydroelectricity. Machinery was now powered by the steam turbine.

Ideas, theories, and inventions emanating from Europe and the United States during the industrial revolution were quickly embraced in Hawaii, where entrepreneurs were constantly looking for better technology to advance sugar production. Dynamite, a powerful tunneling tool, was invented by Alfred Nobel in 1863. In 1876 Samuel Alexander inquired of G. N. Wilcox: "What do you think about 'Fryer's Concreter' for making sugar? Is it all that it is cracked up to be? Is not there some method of clarifying juice better than our use of crude lime?"[9] Sugar planters everywhere profited by a wide sharing of technology through "sugar experiment stations" established around the world. Not all experiments in Hawaii were successful—in 1883, the mongoose was imported from the West Indies to control rats in the cane fields, but it devastated the native ground-nesting birds instead, in a botched attempt at biological control.

Table 4
Plantations and Ditches

Plantation and ditches	Date	Average flow (mgd)*	Capacity (mgd)	Comment
Kauai Plantations				
Lihue Plantation Co./ East Kauai Water Co.		140[†]		Includes Makee
Rice Ditch	1856			
Lower Lihue Ditch				Genesis in Rice Ditch
Upper Lihue Ditch				
Hanamaulu Ditch	ca. 1870			Improved in 1897
Kapaia Ditch				
Waiahi–Kuia Aqueduct (Koloa Ditch)	1915		60–90	Improved in 1926
Waiahi–Iliiliula–N. Wailua ditches	1926			
N. Wailua Ditch		(23)		
Stable Storm				
Hanalei Tunnel	1926	(28)		
Kaapoko Tunnel	1928			
Wailua Ditch		(14)		
Kapahi Tunnel and Makaleha system	1922–1929			
Makee Sugar Co.				
Anahola, Kaneha, Kapaa ditches	ca. 1880–1900			Merged into Lihue Plantation
Grove Farm		26[†]		
Several small ditches	1865–1868			
Halenanahu Ditch	1884			
Huleia Ditch	1893			
Upper Ditch	1917			
Main Ditch (later Lower Ditch)	1928–1948			Genesis in 1865 ditch straightened, lined
Koloa Sugar Co.		20[†]		
Dole's "water lead"	1869			Possibly forerunner of Wilcox Ditch
Wilcox Ditch	1885, 1893			
Mill Ditch	1902			
Waita (Koloa) Reservoir	1906			
McBryde Sugar Co.		95[†]		Includes 50 mgd returned to Wainiha River
Kamooloa Ditch	1906			Built by Grove Farm
Wainiha Powerplant	1906	50	65	Water returned to river

Table 4
Plantations and Ditches

Plantation and ditches	Date	Average flow (mgd)*	Capacity (mgd)	Comment
Pump 3	ca. 1908	34		Includes water pumped from "underground river" (approx. 20 mgd)
Alexander Reservoir	1932	10		
Kilauea Sugar Co.				
System of reservoirs and ditches	ca. 1880–1900			
Reservoirs: Kalihiwai, Stone Dam, Puu Ka Ele, Morita, Waiakalua, and Koloko				
Ditches: Mill, Koolau, Puu Ka Ele, Koloko and Moloaa, Hanalei				
Hawaiian Sugar Co./Olokele Sugar Co.		100†		
Hanapepe Ditch	1891	35	42	
Olokele Ditch	1904	66		
Gay & Robinson				
Koula Ditch Tunnel (Hanonui Tunnel)	1948	40		New intake for Hanapepe Ditch sparked Hanapepe Case
Waimea Sugar Mill Co.				
Waimea (Kikiaola) Ditch	1903	5		
Kekaha Sugar Co.		50†		
Kekaha Ditch	1907	30	40	In 1923, Kekaha Ditch expanded capacity to 50 mgd and average to 35 mgd
Kokee Ditch	1927	15	55	
Oahu Plantations				
Waiahole Irrigation Co./Oahu Sugar Co.		32†		
Waiahole Ditch		42–27	100	Average dropped over time; tunnel capacity: 100 mgd
Waialua Sugar Co.		30†		
Oahu Ditch (Mauka Ditch Tunnel),	1902			Collection and delivery for Lake Wilson Reservoir
Wahiawa, Helemano, Tanada ditches	ca. 1902			
Opaeula Ditch	1903			

Continued

Table 4 (cont.)
Plantations and Ditches

Plantation and ditches	Date	Average flow (mgd)*	Capacity (mgd)	Comment
Kamananui Ditch	1904			
Ito Ditch	1911			
Kahuku Plantation Co.		10[†]		
Punaluu Ditch	ca. 1906	10		
Waimanalo Sugar Co.				
Kailua Ditch				
Maunawili Ditch				
Maui Plantations				
East Maui Irrigation Co.		160[†]	440	
(Old) Hamakua Ditch	1878	(4)		Built by HDC
(Old) Haiku (Spreckels) Ditch	1879			Built by C. Spreckels
Lowrie Ditch (Lowrie Canal)	1900	(37)	60	Built by HC&S/MA
New Hamakua Ditch	1904	(84)		Built by MA
Koolau Ditch	1905	(116)	85	Built by HDC
New Haiku Ditch	1914	25	100	Built by HC&S/EMI
Kauhikoa Ditch	1915	(22)	110	Built by MA
Wailoa Ditch	1923	(170)	160– 195	Built by EMI; originally 160 mgd, later 195
Wailuku Sugar Co.		30[†]		
Waihee (Spreckels) Ditch	1882	10–2	20	Built by C. Spreckels; average is dropping
Waihee (Ditch) Canal	1907	27	50	Average is dropping
Nine other smaller ditches				
Honolua Ranch & Pioneer Mill Co.		50[†]		
Honokohau Ditch	1904	20	35	Developed by Honolua Ranch, now ML&P; replaced by Honolua Ditch
Honolua (Honokohau) Ditch	1913	30–18	50– 70	
Honokowai Ditch	1918	6	50	Replaced 1898 flumes
Kahoma Ditch		3		
Kanaha Ditch		3.8		
Kauaula Ditch		4.5	25.5	Upgraded in 1929

Table 4 (cont.)
Plantations and Ditches

Plantation and ditches	Date	Average flow (mgd)*	Capacity (mgd)	Comment
Launiupoko Ditch		0.8		
Olowalu Ditch		4	11	
Ukumehame Ditch		3	15	
Hawaii Plantations				
Kohala Ditch Co.		30†		
Kohala Ditch	1906			Awini section in 1907
				Wailoa extension in 1908
Kehena Ditch	1914	(6)		
Hamakua Sugar Co./HIC		50†		
Upper Hamakua Ditch	1907	8		
Lower Hamakua Ditch	1910	30	60–45	

*Average flows are based on the historical record except for those in parentheses, which are from USGS records.
†Estimated average total surface water diverted.

The political drama involving the king, Spreckels, Alexander, and Baldwin on Maui was in the context of these changing times of the 1870s and 1880s. The government for the first time wrestled with the problems of water use and development for a new kind of agricultural industry. King Kalakaua acted in a manner consistent with the Reciprocity Treaty he had just executed when he issued a water license to Alexander and Baldwin et al. in 1876. Two years later, Kalakaua gave a water license to Spreckels stipulating that he would acquire the use of water previously assigned to the Hamakua Ditch if that ditch was not completed on schedule. But Alexander and Baldwin did meet the conditions, and Kalakaua honored the contract with them. This was a critical juncture, for when Kalakaua acquiesced to the binding power of the written contract, he relinquished the traditional royal prerogative of changing his mind. He acknowledged and confirmed that contract law, and legislative authority, were superior to royal power in Hawaii. From that time on, the monarchy and later the ensuing provisional and territorial governments readily granted licenses for the use of water to sugar planters, plantations, and their "ditch companies."

5. East Kauai

LIHUE PLANTATION AND EAST KAUAI WATER COMPANY

Lihue Plantation, founded in 1849, was the second-oldest plantation on Kauai (after Koloa Plantation) and one of the oldest in Hawaii. Over the next eighty years the plantation expanded its land base and improved its water system. It acquired Hanamaulu lands in 1870, interest in Koloa Plantation in 1871, and the Makee Plantation in 1933. In 1974 it leased some of Grove Farm's cane lands when that company went out of sugar production. Lihue Plantation was irrigated entirely from gravity flow.

Lihue Plantation developed a water collection system second only to East Maui Irrigation Company. Two entities administered this water system: East Kauai Water Company (EKW) and Lihue Plantation. Altogether there are 51 miles of ditch and eighteen intakes. The system brings down anywhere from 80 to 180 mgd, with an average of 100 to 140 mgd. In 1984, eleven people, seven from EKW and four from Lihue Plantation, were needed to maintain this system.

East Kauai Water Company was established in 1924 with jurisdiction over the waters that arise in and cross state lands. EKW waters are all the flows of the north fork of the Wailua River, the Kapaa and Anahola rivers and their tributaries, and such waters of Hanalei River and Kaapoko stream that are diverted into the North Wailua drainage basin. EKW administers a total length of 34 miles of ditch—the 101 tunnels in the system comprise 11 of the 34 miles. Lihue Plantation manages the rest of the system: the South and North Intake ditches, the Upper and Lower Lihue ditches, the Hanamaulu Ditch, the Kapaia Ditch, and the Waiahi–Iliiliula–North Wailua ditches.

Lihue Plantation had more ditches than ditch records, so only a rough

68

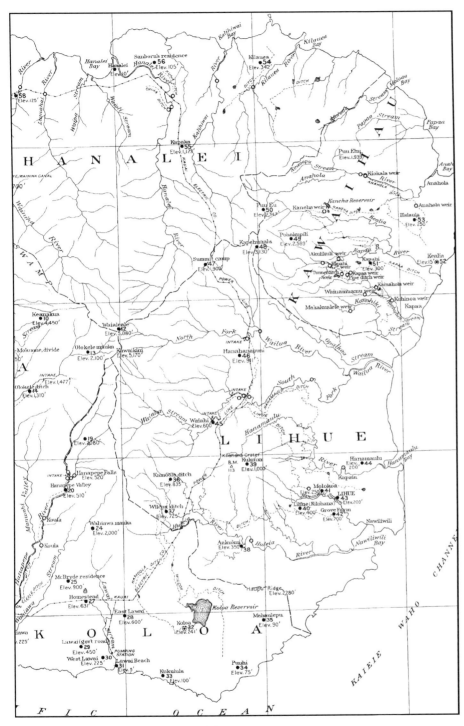

1913 drainage map, East Kauai. (USGS.)

chronology of its water development can be pieced together. It was here that William Harrison Rice built the earliest sugar irrigation ditch in Hawaii, in 1856. This modest ditch was improved by G. N. Wilcox in 1869, then extended to the south fork of the Wailua River in 1876. Enlarged and straightened, it metamorphosed into the Lower Lihue Ditch. In 1895, the north fork of the Wailua River was tapped "by means of a large pipe." Having earlier acquired the Hanamaulu lands, Lihue Plantation contracted G. N. Wilcox and Paul Isenberg sometime around 1897 to improve the Hanamaulu Ditch and install a hydroelectric plant. In 1915, Lihue Plantation constructed the Waiahi–Kuia Aqueduct, known also as the Koloa Ditch, under the supervision of James Robertson. This series of acquisitions and expansions marks the first phase of Lihue Plantation's development.

The second phase took place under the direction of Joseph Hughes Moragne, hired as engineer for Lihue Plantation in 1919, a position he held until 1937. Under his supervision the plantation completed the development of a complex water collection and transfer system, one that spanned and connected several watersheds from Hanalei to Koloa. Before coming to work with Lihue Plantation, Moragne had already established a significant record of achievements in Hawaii.

Born in 1865 in Alabama, Moragne graduated from Auburn Technical Institute (now the University of Alabama) as a civil engineer. He arrived in Hawaii in 1898, worked on the Nuuanu Reservoir on Oahu, ran a lumbering operation in Hilo, and then moved to Kauai in 1905. On Kauai, Moragne's influence was felt from Mana to Kee. His work involved surveys, maps, road construction, tunneling, and many of the irrigation systems. He worked for Gay & Robinson, for Grove Farm, for McBryde and Lihue plantations. He built the 1909 Huleia cane haul bridge, the first reinforced concrete bridge in Hawaii, and the 1917 Upper Ditch for Grove Farm. He designed and installed the water collection system for the Wainiha Powerplant for McBryde Sugar Company. From 1909 to 1919, as the first county engineer for Kauai, he engineered the "belt road" around the island. The winding country road from Hanalei lookout to Haena remains little changed from Moragne's time.

Moragne engineered and supervised many irrigation projects for Lihue Plantation. The Kapahi Tunnel and the Makaleha system were built between 1922 and 1926. In 1926, the Waiahi–Iliiliula–North Wailua ditches were built to bring water from the north fork to the south fork at the 1000 foot elevation. (Koloa Plantation participated in the cost of construction of this project, and in return they received one million gallons of water a day.)

The 1926 Hanalei Tunnel was built by Moragne at a cost of $294,261. This 6028-foot-long tunnel diverted water from the Hanalei River basin at an

Upper Lihue Ditch. Ideally water was moved at the highest elevations possible in order to irrigate *mauka* fields. Here the ditch keeps the water higher than the natural streambed. (Photo: D. Franzen.)

elevation of 1250 feet, through to Wailua River basin, where it dropped into the Maheo stream, a tributary of the north fork of the Wailua River. At the confluence of the Maheo and the south branch of the north fork of the Wailua River, at about the 700 foot elevation, the Stable Storm Ditch transmitted the water westward to a tributary of the south fork, and from there the water could be diverted to the Lihue ditches. Additional water was diverted from the Kaa-

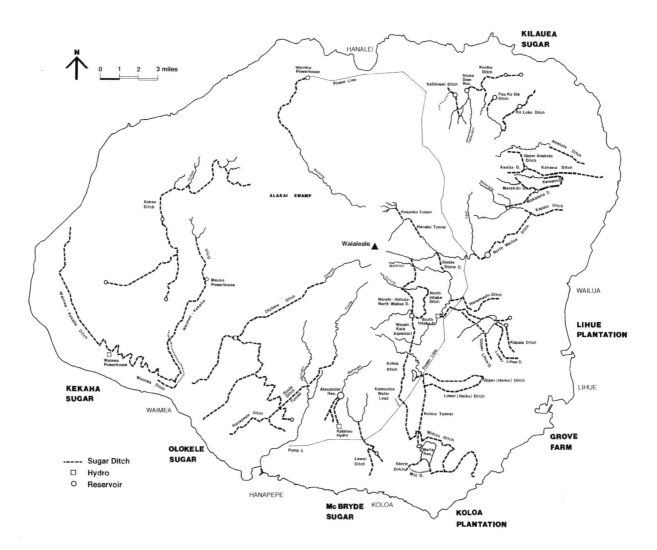

Major sugar plantations and ditches, Kauai.

poko tributary of the Hanalei River into the Hanalei Tunnel in 1928. This tunnel, which cost $150,000, was 3558 feet long and was dug through rock. The project had been foreseen ten years earlier, in 1915, in a report by J. M. Lydgate to William F. Sanborn, manager of Princeville Plantation. In it he said: "It seems very probable that the day may come when the large quantities of water now going steadily to waste on the north side of Kauai, may be conveyed to the South side, where it could be disposed of at lucrative rates."

By 1931, some 79 percent of the 6712 acres of Lihue Plantation's cane land was irrigated by gravity flow. The average water production was 82 mgd. Of its reservoirs the three largest were the Wailua at 242 million gallons, the Upper Kapahi at 30 million gallons, and the Lower Kapahi at 25 million gallons.

Lihue Plantation doubled its size in 1933 with the takeover of Makee Plantation. The Makee Sugar Company was established by James Makee at Kealia, Kauai, in 1877. By the time Lihue Plantation acquired Makee, it had 7200 acres in cane with another 2200 acres planted by independent planters, primarily homesteaders. It had a well-developed water collection and delivery system, too, which delivered an average of some 30 mgd and included Anahola, Kaneha and Kapaa ditches. Altogether it had a total reservoir capacity of 700 million gallons. Lihue Plantation expanded yet again in 1974 when, as noted earlier, it leased Grove Farm's lands.

This trend toward expansion has been reversed in recent times. The Hawaii sugar industry has had to tighten its belt, and Lihue Plantation was no exception. It has cut back on its less productive fields and discontinued maintenance of some of the ditches and reservoirs. In 1994, in a major effort to reduce costs, Amfac/JMB announced the consolidation of many operations shared by Lihue Plantation and Kekaha Sugar Company.

GROVE FARM

George Norton Wilcox, who would one day found Grove Farm, was raised on Kauai by missionary parents. For almost sixty years he was the sole owner of Grove Farm, and he retained control until his death in 1933. Grove Farm was known primarily as a sugar plantation, although it continued to produce dairy and poultry products and diversified crops. George was a boy of seventeen at the time William H. Rice built his ditch in 1856. He earned an engineering certificate at the Sheffield Scientific School at Yale in 1860 and then returned to Kauai.

G. N., as he was called, tried raising sugar in Hanalei, but without success, due to the wet and overcast conditions. Then, in 1864, he was hired by Judge Hermann Widemann, who owned a small plantation in Lihue, to supervise the building of an irrigation ditch. Widemann ignored G. N.'s advice that the ditch was not properly engineered and would not work. G. N. built the project according to specifications, and indeed the ditch did not carry water and the project was abandoned.

In 1864, G. N. first leased and then purchased from Widemann the land that would become known as Grove Farm. It was clear to G. N. that his success in sugar would depend on his ability to irrigate his crop. In January 1865,

G. N. started a new ditch that was finished in three months. While its exact location is unknown, it brought water from the Kilohana slopes to the Puhi area. The ditch cost $556, of which $450 was labor. In 1866, G. N. built another small ditch, which cost $2438.52. Halenanahu Ditch was built in 1884, at a cost of $8564.01, and took eight months to complete. In 1893, Grove Farm built the Huleia Ditch. It was 5.5 miles long and cost $12,375.

George Wilcox turned to various sources for labor for these relatively small ditches. His earliest ditch was built in the native style—that is, the dirt was not lined but was tamped down by the men and possibly by animals as well. It was later estimated by Joseph Moragne that a "good man" would dig probably 5 feet of ditch a day by pick and shovel in the 1860s. Although most of the workers were Hawaiian, historian Bob Krauss in *Grove Farm Plantation* records that two Chinese arrived on the scene in 1865 and completed the tunnel in two weeks for a fee of $14. In 1868 that ditch was extended and expanded, a small job that took ten men three months to complete. Wilcox used Germans and prisoners to build the Halenanahu Ditch: the Germans were paid $17 a month, the prisoners 60 cents a day. He contracted Ahana, who was Chinese, to build the tunnel for $80 and again in 1886, to build the Kokolau Tunnel for $577. Ahana hired Japanese tunnelers for the 1893 Huleia Ditch project. In 1893, G. N. hired prison labor at 50 cents a day, presumably to do the ditching.

At first Wilcox did his own engineering and supervising and set the dynamite, but later he relied on outside engineering and contracted labor. On the 1893 Huleia Ditch, for instance, G. N. contracted engineer Mr. Dove, supervisor John H. Coney, and carpenter Swanston. The modest size of this ditch is suggested by a ten-month material list: forty-eight picks, forty-eight shovels, thirty-six extra pick handles and twelve extra ax handles, a sledgehammer, an alarm clock, tents, eight bars of iron, and, at the very end of the project, four barrels of cement and two of lime. The list also mentions a cook for the "gangs." The Huleia Ditch had twenty-nine tunnels totaling 1.2 miles. Explosives were used for the tunneling, requiring "giant powder," fuses, caps, and sixty-three cases of candles for lighting the tunnels.

In 1906, G. N. negotiated a series of agreements with Koloa Plantation to acquire a right-of-way in exchange for water from Kuia stream. This cleared the way for the Kamooloa Water Lead, started in September and finished in six months, which delivered surplus water from Grove Farm to McBryde Sugar Company. In 1914–1917, Grove Farm built a new "Upper Ditch" on the slopes of Kilohana, engineered by Joseph Moragne, who was assisted by Edward Palmer. By the 1920s Grove Farm had 16 miles of ditches delivering 26 mgd. Starting in 1928, Bill Moragne, Joseph's son and Grove Farm engineer from

Lihue Plantation's lower powerhouse and penstock with cable bridge in background. Almost all of Hawaii's accessible hydroelectric potential was harnessed by the 1930s. (Photo: D. Franzen.)

1928 to 1948, straightened, concreted, and built all new tunnels for the Main Ditch, now called the Lower Ditch, leaving little of the original 1865 ditch.

In 1948 Grove Farm acquired Koloa Plantation and built a vehicular access tunnel from Puhi to Koloa. In 1960, Grove Farm started a water tunnel to Koloa. Due to encountering hard rock and dike water, bad weather, a strike, and poor business conditions, it took five years to complete. Whatever the reasons for the delays, no project in the earlier days took half this long.

Grove Farm's ditch system was a modest one not known for any outstanding technical or physical achievements. This may reflect the limited watershed available to Grove Farm, the small size of that plantation's acreage, or G. N.'s personal sense of scale. In any case, G. N. Wilcox is historically significant for being one of the first to recognize the importance of water development as a prerequisite for a successful sugar plantation.

Grove Farm went out of the sugar business on 1 January 1974. After setting aside some lands for development, it leased its Koloa lands to McBryde Sugar and its Lihue lands to Lihue Plantation for continued sugar production.

KOLOA PLANTATION

Koloa Plantation, established in 1835, is best known as Hawaii's first sugar plantation. Koloa was water poor and as a consequence depended on the surplus water from its neighbors, McBryde, Grove Farm, and Lihue plantations. While long retaining its historic name, it was unable to survive independently. As early as 1871, Lihue Plantation had bought a half-interest in Koloa Plantation.

The earliest irrigation at Koloa Plantation was in 1869, a drought year. Manager George Dole and George Charman dug a "water lead" to save the cane during the drought. This was most likely the infancy of the Wilcox Ditch, which was extended in 1885, bringing nearly 200 more acres under irrigation, and again in 1893, with the help of a 3000-foot siphon, to irrigate another 400 acres of the inland Puuhi fields. The capacity of the Wilcox Ditch was 8 mgd; the average daily flow was probably half that. The Wilcox Ditch supplied all of Koloa Plantation's irrigation water until 1897, when small wells were drilled at Mahaulepu. These were marginally successful, and subsequent tunnels and wells produced little additional water. Mill Ditch was built, in 1902, for sending water to the factory.

Constrained by lack of surface and groundwater sources, Koloa concentrated on developing water storage. The Koloa Marsh was an ideal location for a reservoir. With little initial investment and subsequent upgrading, Koloa Plantation developed the second-largest reservoir in Hawaii. With a 2.3-billion-gallon capacity, the Waita (Koloa) Reservoir is Koloa Plantation's most notable irrigation achievement.

The seeds for this project were sown in 1904, when company president Hans Isenberg contracted J. S. Malony to survey the marshlands outside of Koloa town for a 95-acre reservoir. In his report, Malony recommended that the entire marshland be put in reservoir, one that would cover 255 acres. Koloa Plantation, ignoring this advice, built the smaller project, known variously as the M&M or the Hauiki Reservoir. It proved such a success that the expansion previously recommended by Malony was immediately approved, and work began on the larger Marsh Reservoir.

Marsh Reservoir, with a capacity of 1 billion gallons, was completed in 1906 for approximately $30,000. It was separated by a dam from the M&M Reservoir. In 1908 its retaining dam was raised, increasing capacity to 2.1 billion gallons, at a cost of about $17,000. Over the years the Marsh Reservoir dam settled and eroded, so in 1931 the dam was raised 3 feet and reinforced with rock, once again increasing the capacity, this time to 2.3 billion gallons. The reservoir now covered 525 acres of land, submerging the old dam dividing the original Hauiki Reservoir and the Marsh Reservoir. This separator dam was then raised 5 feet and a railroad track was placed on it for access to the back fields. Known for years as the Koloa Reservoir,[1] it is now known as the Waita Reservoir. Despite the lack of a source of surface water, Koloa Plantation's storage system enabled it to irrigate over 70 percent of its fields from mountain sources.

The water source for the Koloa Reservoir was supplied through the modest Wilcox Ditch. When this ditch proved inadequate to keep the reservoir supplied, an agreement was reached in 1914 between Koloa Sugar Company, Lihue Plantation, and G. N. Wilcox whereby Lihue Plantation would build the Waiahi–Kuia Aqueduct, also known as the Koloa Ditch. Lihue Plantation took water from Waiahi and Kuia streams over Grove Farm lands into Koloa's Wilcox Ditch and the Waita (Koloa) Reservoir. Built from plans prepared by Jorgen Jorgensen and supervised by James L. Robertson, it involved a total length of 3.3 miles: 14,685 feet of tunnel (the longest was 5845 feet), 467 feet of flume, and 2320 feet of ditch. Although the capacity was projected at 100 mgd, in fact it was more like 65. In 1926, Lihue Plantation enlarged the capacity to 90 mgd to better capture floodwaters.

James L. Robertson, one of those engineers who worked on numerous projects in Hawaii, arrived in 1897 to observe the ceremony marking the annexation of Hawaii to the United States. Like so many others, he came to visit and ended up making Hawaii his home. Although he specialized as a "builder of roads," Robertson contributed substantially to the development of irrigation projects. Besides the Koloa Ditch and reservoirs for McBryde, he worked on the Hilo Railroad, Wainiha Powerplant, and the Kekaha Ditch.

In 1948, Koloa Plantation, along with its mill, was acquired by Grove

Farm. In 1965, Grove Farm built a tunnel to bring the waters from Kuia directly into the Waita Reservoir. Grove Farm leased these cane lands to McBryde Sugar Company when it terminated sugar operations in 1974.

MCBRYDE SUGAR COMPANY

Benjamin F. Dillingham combined the McBryde Estate (which itself was a combination of Koloa Agricultural Company and Wahiawa Ranch) and Eleele Plantation to form McBryde Sugar Company in 1899. McBryde Sugar had

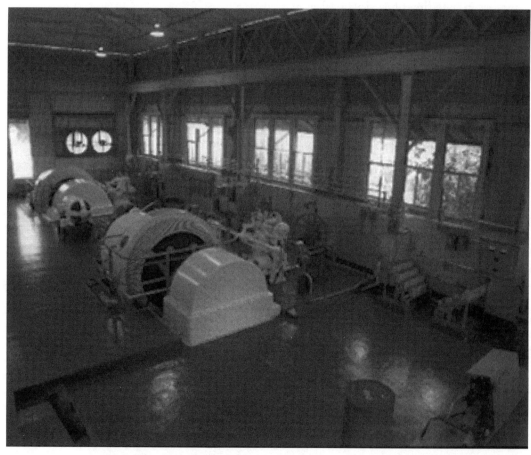

McBryde's Wainiha Powerplant. The Wainiha Powerplant is capable of generating 4000 kilowatts. It continues to produce more energy annually than any other hydro-electric plant in Hawaii. The plant was damaged by Hurricane Iniki on 11 September 1992; it was placed back on line in June 1993. (Photo: D. Franzen.)

20,000 acres in Kalaheo, Hanapepe, Eleele, Lawai, and Koloa, including an *ahupuaʻa* that reached to the top of Mount Kahili. Despite this, the new company did not have access to a sufficient supply of surface water, so it concentrated on pumping groundwater and water storage. By 1903 the company had developed a fairly extensive system of reservoirs with a combined capacity of 800 million gallons.

When McBryde's cost of running its coal-burning steam pumps proved prohibitive, the company turned to cheaper energy sources—specifically hydroelectric power and burning bagasse for fuel. The center of the power grid was at Pump 3, which was further distinguished by having intercepted an underground river. The first of McBryde's two powerplants to go on line was built at Wainiha in 1906.

McBryde's Wainiha Powerplant is the earliest hydroelectric powerplant of any significant size built in Hawaii—and to this day remains the largest in annual power production. Built in Wainiha, on the northern, windward side of Kauai, its primary purpose was to provide power to McBryde's pumps on the leeward side, although it generated a great deal more power than that. Since Wainiha was far removed from McBryde sugarcane, or from any cane at all for that matter, water was returned to the river after leaving the powerhouse.

Technically speaking, Kauai Electric Company built the Wainiha Powerplant. This company was incorporated around 1904 as a subsidiary of McBryde specifically to provide the sugar company with cheap energy, somewhat akin to the water companies set up by other plantations. It was later unincorporated and became an integral part of McBryde. In 1930 a new Kauai Electric Company was formed, the forerunner of the present public utility company of that name. McBryde Sugar Company was the majority stockholder of the newly formed company.

The preliminary surveys for the water collection system and the plant location were taken in late 1904, and the plant was completed in August 1906. James Robertson installed the power station and power lines, with the help of Alfred Menefoglio, who remained on as supervisor for many years. Alonzo Gartley, manager of Hawaiian Electric and a fine amateur photographer, helped install the equipment.

The primary water source was the Wainiha River, with the intake at the 700-foot elevation. Seventeen tributaries and the Maunahena (Mauna Hina) stream were diverted between the intake and the plant. The water collection system was engineered by Joseph Moragne and contracted by Henry A. Jaeger. A wharf was built on the beach at Wainiha, and warehouses erected near it. A light railroad ran up to the work site, where the camp was located. The ditch included thirty-two tunnels, with a total length of 17,400 feet, and eight ditches with a total length of 5600 feet. Water was received into a concrete-lined forebay (pool not affected by surges) and then dropped through a 1612-foot-long penstock (a pressurizing pipe) with a head of 565 feet. Ditch capacity was 65 mgd and its low flow, based on river gauging, was 20 mgd. (At present capabilities, median flow at the powerhouse, calculated from power production, is 50 mgd.)

At completion the Wainiha Powerplant generated a higher voltage than any other plant west of the Rockies: it was able to transmit 33,000 volts at 25 cycles. It had two 1200-kilowatt generators, Pelton waterwheels, two 70-kilowatt exciters, seven 500-kilovolt-ampere transformers (two banks of three transformers each with a seventh spare), and a switchboard. At some point a third generator unit, dated 1915, was installed. McBryde modified the Wainiha plant several times, increasing its capacity to 57,000 volts at 60 cycles.

Once the powerplant was completed, power was transmitted the 35 miles to Hanapepe by means of a line that crossed Wainiha, Lumahai, and Hanalei valleys, traveled up the ridge back of Kalihiwai to the mountain divide between Kalihiwai and Wailua, headed toward Lihue, passed between Haiku and Lawai, and ended up at Hanapepe. The power was originally fed to Pump 2 but went to Pump 3 when it was built in 1908. Aluminum wire, new and somewhat experimental, was used instead of copper. Its resistance to corrosion and its comparatively light weight recommended it for this job. Permanent trails and access roads were built for construction and maintenance of the powerline. With the addition of this powerplant to its facilities, McBryde Sugar Company was able to convert its pumping system to electricity and realized a substantial savings in fuel bills.

In 1928, McBryde built a second, 1100-kilowatt hydroelectric powerplant at Kalaheo, using water stored behind the Alexander Dam. In the early years, McBryde directed all the energy generated by the two hydroelectric plants and a bagasse-fueled plant at the mill to Pump 3 and distributed power from there. Seventy percent of the power was used for pumping. After servicing plantation needs, excess power was sold, over the years, to Kauai Railway Company, Kauai Pineapple Company, Kilauea Sugar Company, and Kauai Electric Company. Wainiha Powerplant and the Kalaheo Powerhouse are monitored and regulated from Pump 3.

Pump 3 was one of four pump stations that tapped both in and under the Hanapepe River. (Three of them have since been abandoned.) Pump 3 was uniquely successful in water production. At Pump 3 a vertical shaft descends 90 feet to the pump room and sump. Forty feet below that a network of skimming tunnels was built starting in 1908. In 1927 one of these tunnels was being extended when, according to field notes, it "blew out a hole in the roof [of a lava tube 4 feet thick] and got a stream +/– 2 mgd, under heavy pressure"[1] The tunnel had in essence intercepted an underground river. Up to 32 mgd was realized from this source at times, though the average was closer to 15 or 20 mgd. In addition, Pump 3 is recharged with surface water diverted from the Hanapepe River. Of the 34 mgd consistently drawn from Pump 3, some was put into Farmers Ditch for sugar and taro farmers along the river, but most of it was pumped up to Pump 3 Ditch for distribution to the fields.

While McBryde was more dependent on groundwater than any other plantation on Kauai, it did have access to the Wahiawa watershed, which includes the Kanaele Swamp. Determined to build a reservoir to capture the runoff from this watershed, McBryde started the Alexander Reservoir in August 1928 under the supervision of engineer Joel B. Cox. By this time, although all the major surface-water collection systems had been completed, very few large reservoirs had been built. In fact, conditions for water storage were not generally ideal in Hawaii. The conditions described on Maui by geologist H. I. Stearns could be applied to all the islands: "Satisfactory sites for reservoirs do not exist. The rocks are so permeable that all unlined reservoirs leak heavily and are used chiefly for overnight and storm-water storage. Also, the steep gradients of the canyon floors and the great amount of debris carried by the streams are added difficulties in building and maintaining satisfactory large reservoirs."[2] The only Hawaii plantations that developed any substantial water storage capacity were McBryde and Koloa plantations on Kauai and Waialua Sugar Company on Oahu.

The Alexander dam was 120 feet high, 620 feet long at the crest, and 640 feet thick at the base. It was built by hydraulic fill: embankment material was sluiced to the site. Work was done in three eight-hour shifts of sixteen men each who worked by floodlight at night. On 26 March 1930, when on the verge of completion, the dam broke under the impact of torrential rains. The resulting mudslide took six lives and injured two more. The dam was rebuilt and finished in December 1932 at a cost of $2.2 million, which included the Kalaheo Powerplant. Although initially estimated at a capacity of 700 million gallons, it actually has a capacity of over 800 million gallons, one of Hawaii's largest. Moreover, the reservoir boasted the second-highest earthen dam in Hawaii.

Controversy was sparked on the banks of Omao stream in Koloa in 1906, when McBryde Sugar Company expanded an existing *mauka* ditch and reservoir on Omao stream, depriving downstream user Koloa Plantation of its usual supply. In return, Koloa Plantation built two dams, taking all the water from Omao stream and its easterly branch. McBryde workers promptly removed them in the early morning hours. Koloa workers tried to replace the dams but were deterred by armed McBryde forces. They then blasted the lower

Alexander Reservoir. The 1932 Alexander Reservoir dam was the second-highest earthen dam in Hawaii, at 120 feet high, 620 feet long, and 640 feet thick at the base. Earthen material was excavated from nearby ridges and sluiced to the site. The dam was faced with rubble and then finished with fitted rock. (Private collection.)

McBryde dam. When the McBryde men attempted to repair this they were barraged with shovels of earth from the bank. And on it went for three years, firearms displayed and attorneys kept busy. The matter was resolved in 1909 by mutual agreement.

Above and beyond its considerable technical achievements, McBryde is known for its involvement in one of the most sweeping Hawaii Supreme Court decisions of the twentieth century: the 1973 McBryde Decision (discussed in Part I).

Alexander & Baldwin acquired control of McBryde in 1909. In 1974, McBryde expanded to take over Grove Farm's Koloa cane lands when that company went out of sugar production. Two decades later, McBryde itself went out of business, one of six sugar companies that fell to the economic pressures of 1994–1995.

KILAUEA SUGAR COMPANY

Kilauea Sugar Company, on Kauai's north shore, had to make the best of marginal conditions. Plagued by rocky terrain, small size, few water resources, and its remote, windward location, it never enjoyed the success of other, better-situated plantations. Kilauea Sugar planted its first fields in 1877, was incorporated in 1880, and crystallized its first sugar that same year. Kilauea became known for innovations in field practices and cost management.

Although Kilauea Plantation's gravity flow system was not especially early or technically innovative, it is a good illustration of what a small system, using modest, unlined ditches and reservoirs, can accomplish. It was nicely interconnected and—using a network of reservoirs with a total capacity of over 730 million gallons—was able to irrigate some 3000 acres of cane stretching over 6 miles.

Kilauea Plantation's four small ditch systems connected six reservoirs: Kalihiwai, Stone Dam, Puu Ka Ele, Morita, Waiakalua, and Koloko. Koloko, with a 408,856-gallon capacity, was the largest. All but Kalihiwai were built by 1911. The easternmost reservoir was the Kalihiwai, also known as Drinking Water Reservoir, which provided domestic water for Kilauea town through an 8-inch pipe. The water from the Kalihiwai Reservoir went to Stone Dam.

Stone Dam Reservoir was created by blocking the streambed just below the convergence of Pohakuhono and Halaulani streams. It was supplemented by water from the Kalihiwai and Hanalei ditches. The Kalihiwai Ditch was slightly over a mile long and collected water from three tributaries of the Pohakuhono stream. The 1922 Hanalei Ditch diverted water from the Kalihiwai River and had a capacity of 10 to 15 mgd. At 3.8 miles it was the longest collection ditch in Kilauea's system. From Stone Dam, water was sent in two directions. It went east 3 miles to Kilauea town, by way of Mill Ditch, and west passing through the Waiakalua Reservoir by way of the 8-mile Koolau Ditch. Puu Ka Ele stream was tapped by the ditch of the same name and the water was stored at the Puu Ka Ele Reservoir. The ditch continued on to the Morita and Waiakalua reservoirs. Koloko Reservoir was fed by the Koloko and Moloaa ditches. From this reservoir the water went to the fields as well as to the Waiakalua Reservoir.

By 1931, Kilauea Sugar had a total of 3875 acres in cane, three-quarters of which was irrigated. There was 33 miles of main ditching, practically all of it earth-lined, the balance stone-lined. It is not hard to imagine the plantation manager's enthusiasm when he was able to report a new ditching plow. L. David Larsen reported to the directors in 1922: "Although a crude home-made affair, it illustrates the possibilities of implements replacing hand labor for much of our ditch work, in the fields as well as in the big ditches."

In 1971, when Kilauea Plantation closed, local and state government, along with agent C. Brewer and individual farmers, began to search for alternative agricultural options. Besides cattle, which had long been an industry in this area, papaya, guava, prawn ventures, and agricultural subdivisions were established with the expectation of being able to use the Kilauea water system. However, no mechanism was established to secure the easements or maintain the old system. Over the years the connections between reservoirs and delivery systems were destroyed by roads, pasture, development, neglect, and intent. The Hanalei Ditch was abandoned, its flumes and siphon no longer operable. The connection from the Kalihiwai Reservoir to Stone Dam was destroyed, as was that between Puu Ka Ele and Morita reservoirs. Puu Ka Ele and Koloko reservoirs' delivery systems were gone. C. Brewer established Kilauea Irrigation Company, a public utility, to administer the surviving sections that service its guava farming operation. By the mid-1990s, some reservoirs stood alone with little utilitarian purpose.

6. West Kauai

HAWAIIAN SUGAR COMPANY (MAKAWELI PLANTATION)

The Hawaiian Sugar Company was incorporated in 1889 with Henry P. Baldwin as principal shareholder. The company, better known locally as Makaweli Plantation, signed a fifty-year lease with Gay & Robinson for land on the west side of Kauai. After building a new mill at Kaumakani, Baldwin turned his attention to getting water from both the Olokele and Hanapepe rivers, both large watercourses by Hawaii standards. This undertaking, he estimated, would cost half a million dollars.

The Hanapepe Ditch was designed by Baldwin himself, who despite his lack of formal training had demonstrated his considerable ability on the Hamakua Ditch on Maui. His work was reviewed by G. F. Allardt, a consulting engineer from San Francisco, who could find no fault with it.

Construction commenced on the Hanapepe Ditch in 1889 under the supervision of B. C. Perry and was completed on 25 April 1891 at a cost of $152,013. The ditch ended in a reservoir established in an ancient volcanic crater, 900 feet across, which had been enlarged to a depth of 30 feet. In the 13.5 miles from the intake on the Hanapepe River to the reservoir, there were 5570 feet of pipe, 12,300 feet of flumes, 1017 feet of tunnels, six siphons, and 16,500 feet of ditch. The longest tunnel was 373 feet; the longest siphon, 1923 feet. The pipe—riveted steel of 40-inch diameter coated inside and out with asphalt and coal tar—was given a fifty-year life expectancy. The flumes followed the canyon wall along the east side of the Hanapepe Valley; then the water crossed the river four times by pipe. The final crossing was at an elevation of about 440 feet, at which point the water was brought across the valley and delivered to the plantation via a siphon. The last 6.5 miles was dirt ditch

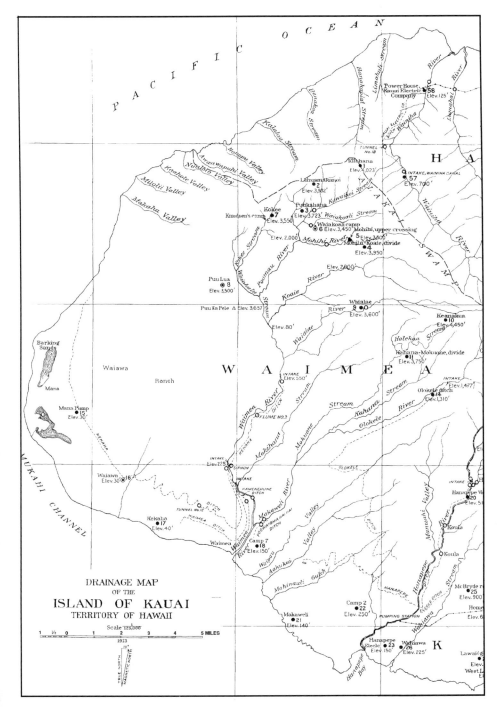

1913 drainage map, West Kauai. (USGS.)

"cheaply done with the plow and scraper."[1] The capacity of the Hanapepe Ditch was variously reported at 42 and 51 mgd with an average of 35 mgd. With this new source, Baldwin was able to irrigate 3000 acres of cane.

But Baldwin envisioned a much larger plantation. He proposed to expand by opening more *mauka* lands to irrigation by tapping into the upper reaches of the Olokele River, on land owned by Gay & Robinson. Agreements for the water and lands were drawn up in 1900. Not all of these agreements were actually executed by Gay & Robinson, but Hawaiian Sugar Company proceeded

Opening of Olokele Ditch, 1904. Michael O'Shaughnessy engineered the Olokele Ditch, which for the first time in Hawaii replaced open ditches built on the contour with long tunnels. (Photo: A. Gartley. Private collection.)

with the improvements anyway—a decision that would prove to be a source of trouble between the two companies in the years ahead.

The Olokele Ditch, engineered by Michael M. O'Shaughnessy, revolutionized ditch building in Hawaii by introducing long tunnels to replace open ditches built on the contour. The Olokele Ditch was started in 1902 and completed in 1904 at a cost of $360,000. It included 8 miles of 7-by-7-foot tunnels and 5 miles of ditch. It delivered water at the 1075 foot elevation. With an average flow of 66 mgd and a minimum of 46 mgd, it was noted for its efficiency and low maintenance: only five ditchmen were required for its upkeep. With the completion of the Olokele Ditch, the Hawaiian Sugar Company had one of the best gravity supplies in the islands.

With good soil, a consistent and plentiful supply of water, and year-round sun, the conditions were ideal for sugar. The average annual production at Makaweli Plantation went from 12,000 tons to 20,500 tons. The $500,000 bond issued to cover the new improvements was redeemed by 1909. Makaweli Plantation was one of the most successful, and most profitable, in Hawaii. And yet by 1923 manager Ben Baldwin wished to increase the water supply by 50 percent.

Gay & Robinson continued in an uneasy relationship with Hawaiian Sugar Company and its agent A&B for many years. When it came time to renegotiate the leases, they were unable to reach agreement. Hawaiian Sugar Company was terminated in 1940 and a new company, Olokele Sugar Company, was organized under the agency of C. Brewer & Company. It was physically the same plantation, but the control and management were entirely new.

Olokele Sugar Company signed new leases with Gay & Robinson. But there was a significant difference between the old and new leases: the disposition of the water collection systems. The earlier lease had allowed Hawaiian Sugar Company to develop and maintain its water collection system. Under the new lease, Gay & Robinson assumed complete control of these systems and delivered the water to Olokele Sugar Company. Thus perished the Hawaiian Sugar Company—to some extent, at least, a victim of water rights concerns.

By the time the new Olokele Sugar Company was started, the Hanapepe Ditch leakage was estimated at possibly a million gallons a day. To remedy this, in 1948 Gay & Robinson and C. Brewer built the Koula Ditch Tunnel to replace the Hanapepe Ditch flumes. Referred to locally as the Hanonui Tunnel, Koula Tunnel is actually two tunnels, one 13,000 feet and the other 4000 feet, totaling over 3 miles.

If there is such a thing as too much success, the Koula Tunnel is surely an example. For many decades, both upstream user Makaweli Plantation (and then Olokele Sugar Company) and downstream user McBryde Sugar between them diverted essentially all the water from the Hanapepe River, so that the mouth of the river was usually dry. That balance—the allocation of that water between the two plantations—was upset with the construction of Koula Tunnel. By taking more water at the upstream diversion, the tunnel diminished the supply previously available to downstream user McBryde. McBryde sued Gay & Robinson. The Hanapepe Case resulted in the landmark 1973 Supreme Court decision in water rights commonly known as the McBryde Decision.

In 1994, Gay & Robinson announced that it was taking over Olokele Sugar Company—a surprising case of the minnow swallowing the whale that leads one to contemplate the power of controlling the water source.

WAIMEA SUGAR MILL COMPANY

In 1884, W. D. Schmidt, E. E. Conant, and G. B. Rowell, along with several other investors, leased land from the family of missionary George Rowell and incorporated the Waimea Sugar Mill Company. W. D. Schmidt was the majority stockholder and first manager. The Waimea Sugar Mill Company, with only 275 acres in cane by 1890, was a small one.

The Waimea Sugar Mill Company was irrigated with groundwater and water pumped from the swamps, which tended to be brackish. Excessive use of the groundwater, however, soon caused the water to become more saline. The shareholders' minutes of 25 February 1902 warned: "The obtaining of fresh water from the Waimea River seems to be the only plan of saving the plantation and continuing it, owing to the condition of its present water supply."

With assets of $41,000 and indebtedness to Castle & Cooke of $47,000, it would appear that Waimea Sugar Mill Company was not in a position to undertake a major development project. Nevertheless, it negotiated water rights with Gay & Robinson for $800 a year, as well as a right-of-way to the Waimea River for another $400. Engineer E. Tappan Tannatt estimated the cost of obtaining water from the Waimea River at $29,000. By December 1902, surveying for the new ditch had begun. The Waimea Ditch, later known as Kikiaola Ditch, was completed in September 1903 at a cost of $36,278. There were no tunnels along its almost 6-mile length. It delivered 5 mgd from the Waimea River to this small plantation.

Hans Peter Faye, Knudsen's nephew and manager of Kekaha Sugar Company, bought the Waimea Sugar Mill Company lands from the Rowells in 1904. Faye's interest in acquiring this small plantation was primarily for the right-of-way it had acquired from Gay & Robinson. As the acquisition of this right-of-way would seem to have been to Kekaha Sugar Company's advantage, it is curious that when Faye offered his holdings to Kekaha Sugar it was rejected. Thus Faye became owner of one plantation and manager of the other.

Although working in the water as pictured here might have been refreshing at times, this was not always the case. Occasionally, enormous flows were released into tunnels— as during the digging of the Waiahole Tunnel. The engineer reported "for the last three hundred feet the tunnel was a whirling rainstorm, a giant shower bath, a water-spout and a typhoon combined. From sides, roof and face the water spurted in continu-ous streams, in some instances with a force against which one could not stand." (Private collection.)

Faye was a gifted sugar planter: by 1910 he had paid off Waimea Sugar Mill Company's debts and was paying dividends. The freshwater supply from the Kikiaola Ditch made the critical difference. By 1910, this plantation was producing 7.25 tons per acre, a most respectable yield for the time. By 1915 Faye was sole owner of all 5000 shares of this company.

From 1914 to 1922, the manager of Waimea Sugar Mill Company was George R. Ewart, a Scot. In 1923 Ewart, an engineer, repaired and realigned Kikiaola Ditch and built tunnels to replace the iron flumes. Upholding the Scottish reputation for assiduous thrift, Ewart wrote that these repairs cost "something like $45,471.30."

In 1969 the Waimea Sugar Mill Company reorganized into the Kikiaola Land Company, leasing its ditch and cane lands to Kekaha Sugar. As Kekaha Sugar had its own ditch taking water from the Waimea River, it closed the Kikiaola Ditch from the source to the town, putting water back into the low-land section of the ditch for field irrigation.

KEKAHA SUGAR COMPANY

The lands of Waimea and Mana, on the west side of Kauai, were once vast marshlands. There were large villages of Hawaiians living at Barking Sands, and rows of thatched houses along the foothills. Hawaiians traveled by canoe from Mana to Waimea by way of an inland waterway. Although Hawaiians at some earlier time had attempted a ditch at Waiele in order to drain the marsh, they apparently abandoned the project when they struck sandstone.

Valdemar Knudsen started what was to become Kekaha Sugar Company in 1878. He held crown leases on Kekaha, Kokee, and Mana lands, marshy land that he initially sublet to small sugar and rice farmers. Noting the success of the fledging sugar industry elsewhere in Hawaii, he determined to try sugar in Kekaha. Knudsen widened and deepened the old Hawaiian ditch with pick and shovel, and by this means drained the marsh. He and Captain Hans L'Orange then planted sugarcane, using groundwater for irrigation.

Knowing that the groundwater supply was limited, Knudsen looked to the Waimea River. Considering the eventual success of this river as a water source, it is somewhat surprising that time and again the experts concluded that tapping the Waimea was not a viable option. The first of several feasibility studies was conducted in 1881 by G. N. Wilcox, though his recommendations do not survive. The second study, by an unidentified engineer in 1892, advised: "I am fully convinced that it is impossible to obtain a water supply from the Waimea River at a high altitude except at a cost which would be prohibitory as a business investment." Even at a lower elevation a ditch would cost $160,000 and a great deal in maintenance. "I cannot think," said this engineer, "that the scheme is likely to prove a financial solution of the water supply difficulty at Kekaha and certainly cannot recommend it as such." A third study by Jim Taylor in 1898 concluded yet again that such a project would be too costly.

Kekaha Sugar Company, Waimea Sugar Mill Company, and the small, independent rice and sugar growers in the area intensified their use of groundwater. In 1898, Kekaha Sugar installed a new 10-mgd-capacity Fraser and Chalmers pump. In the early 1900s, drought and overuse of groundwater ruined that supply for everyone. Within a few years the salt content had gone from 7 grains to 100 grains per gallon. The Mana well levels dropped 4 feet in four years. Although the plantations also had to contend with leaf hoppers, extremely hot weather, and a shortage of labor, their major problem was lack of water. The supply dropped from 23 mgd in 1904 to 17 mgd in 1905. Pumping one well caused others to fail. Crops were lost or simply not planted at all. Kekaha Sugar Company was forced to take a much more serious look at developing Waimea River water.

Kekaha Sugar Company had incorporated in 1898 with Hans Peter Faye as the new manager. Faye dramatically influenced the sugar industry during his forty-five years on the west side. He was born in Drammen, Norway, in 1859, came to Hawaii in 1880, and arrived on Kauai in 1882. Hans L'Orange was his brother-in-law, and Valdemar Knudsen his uncle. He started his career in sugar on land leased from Knudsen in 1884, and later purchased Rowell's Waimea lands, the Waimea Dairy, and, in 1906, gained control of the Waimea Sugar Mill Company. As manager of Kekaha Sugar Company for thirty years, he oversaw the construction of the Kekaha and Kokee ditches, supervised the reclamation of the marshlands, and is credited with major innovations in field practices.

Faye, impressed with the difference the fresh water supplied by the new Waimea (Kikiaola) Ditch was making at Waimea Sugar Mill Plantation, contracted yet another study. He hired engineer J. S. Malony to investigate the possibility of using the surplus water from the Kikiaola Ditch. Malony determined the supply insufficient, so he proceeded to survey the Waimea Valley at three different levels and concluded that a new ditch could be constructed at the 500-foot elevation at a cost of $130,000. This time it looked like a reasonable expense.

Started in May 1906 and finished in September 1907—and with the help of a new hydroelectric plant—the Kekaha Ditch was an immediate success. (This ditch was originally referred to as the Waimea Ditch and is variously known as the Waimea–Kekaha Ditch and the Kekaha Ditch, the name used here. The 1903 ditch built by Waimea Sugar Mill, also originally known as the Waimea Ditch, is here called the Kikiaola Ditch.) Once leases were arranged with the government and other Waimea landowners, the Kekaha Ditch was begun. The construction contract went to J. S. Malony, who hired James L. Robertson as supervisor. Originally the ditch was 20 miles long—16 miles on the *mauka* lands and 4 on the lowlands—and it was later extended another 8 miles. Water was taken from the Waimea River at an elevation of 550 feet. Most of the unlined ditches and tunnels were driven through hard rock and were unlined. A 2190-foot steel inverted siphon, since replaced, crossed the Waimea River.

Although the construction cost was estimated at $130,000, in fact it reached $240,000 to $290,000, not including the powerplant, a later extension, or repairs. The capacity was rated at 45 mgd, and average flow was 30 mgd. Four to five hundred additional acres above the ditch were put into cane, utilizing the hydropower to pump the water to the higher elevation. It was noted that the tunnels would have to be cemented if more land was put into irrigation. At this point the groundwater pumps, although kept in working order, were retired.

Ditches reached up into the Waimea mountains. While the sides were reinforced with cut stone to minimize erosion, the ditch floor was unlined. (Private collection.)

Labor raiding occurred to one degree or another on many plantations throughout Hawaii, although it was strongly discouraged by the sugar factors. It is within the context of building the Kekaha Ditch that we glimpse the beginning of such a struggle between Makaweli Plantation (Hawaiian Sugar Company) and neighboring Kekaha Sugar. Through correspondence between their respective managers, Ben D. Baldwin and Hans P. Faye, we learn that a struggle was well under way in 1905 and continued into the 1920s. In 1905, Faye complained to Hackfeld & Company, agents for Kekaha Sugar, that a great many of his men were leaving to take up favorable contracts at Makaweli. Since the Olokele Ditch was finished by then, this additional labor must have been used to expand the plantation's fields.

In 1906, Baldwin complained to his agent, Alexander & Baldwin, about the exodus of workers to Kekaha Sugar Company, which was then embarking on the construction of the Kekaha Ditch. A&B filed a forceful complaint with Kekaha Sugar's agent, Hackfeld & Company, which it relayed to Faye: "We have already advised you that Mssrs Alexander & Baldwin Ltd DID lodge a strong complaint with us in regard to your taking away men from the Makaweli Plantation and it appears that Manager Baldwin has greatly exaggerated the matter."[1]

Faye assured Hackfeld that he expected "to secure most of the laborers for constructing the new ditch without interfering with the other plantations." So Baldwin wrote directly to Faye:

You state that you do not want our men. Of course contractors will claim that men come from Honolulu, Hawaii, Waialua, G——, etc. and to a certain extent their claim is true, but when we sift this question down to the bottom we generally find that a great proportion of the men are taken from right under our noses here, however, I will not say anything further at this writing in regard to this subject, as you have assured me that you will prevent as far as possible the taking of our men.

But things did not improve, and later Baldwin again complained emphatically to Faye: "You stated among other things that it was *not so* that your contractors were invading our camps to entice our men away. It *was so* and they are still *doing it*." Baldwin writes that although he has heard that Kekaha has thirty dissatisfied loaders who are threatening to come over to Makaweli to work, of course he will not hire them. The record then falls silent.

The Kekaha Ditch was not enough. By 1909 the *mauka* cane lands had been so expanded that the plantation again relied on its groundwater pumps to supply water to irrigate the *makai* lands. The water in those wells was reported "much improved" from its previously saline condition. But Kekaha Ditch's water production had dropped so dramatically that by 1912 the water from the ditch was inadequate even to run the hydroelectric plant. The cause was increased seepage. It was then decided to develop more water by extending the upper portion of the ditch another 280 feet above the existing intake and, at the same time, to build a second hydro, the Mauka Powerhouse, at the original intake.

Although a large storm in 1920 inflicted substantial damage on the Kekaha Ditch and Mauka Powerhouse, the plantation was hesitant to undertake expensive maintenance until the government, which held the lease to nearly all of Kekaha's sugar lands, renegotiated new lease agreements. These were executed in 1923. They stipulated that 2000 more acres of *mauka* pasture land should be put into cultivation and allowed water development in the head reaches of Waimea Canyon to irrigate this additional acreage. In 1923, the company undertook major repairs of the Kekaha Ditch, expanding its capacity to 50 mgd and its average to 35 mgd. At the same time Kekaha Sugar started construction on a new ditch, to be known as the Kokee Ditch.

Faye started the Kokee Ditch—sometimes referred to as the Great Mauka Ditch—in 1923. The construction contract went to George Ewart, previously with Waimea Sugar Mill Company. Although the cost estimates were between $200,000 and $250,000, the actual cost was variously reported as $500,000 and even $680,000—the higher amount might have included the hydro plant.

This ditch diverted tributaries of the Waimea River in the Kokee area—starting at over 3000 feet elevation with the Mohihi and including the Waiakoali, Kawaikoi, Kauaikinana, and Kokee streams—and comprised forty-eight tunnels averaging 1000 feet, the longest being 3000 feet. The total length was 7 miles of tunnel and 12 miles of open ditch, measured to Kitano Reservoir. Water was running through the ditch by January 1925, and the final upper section of Mohihi was completed early the next year. Puu Lua Reservoir, the major storage facility for this system, was finished in 1927, with a 262-million-gallon capacity, at a cost of $168,581. The capacity of the ditch is still 55 mgd up to the reservoir (beyond that point it is 26 mgd); the average flow is 15 mgd.

The average delivery of the Kekaha and Kokee ditches combined is 50 mgd. Besides its surface water sources, Kekaha Sugar has four pumps with a capacity of 26.5 mgd and an average of 22 mgd. The Huluhulunui Shaft, which pumps 10 mgd, provides the water to run the factory.

Kekaha Sugar planted on a variety of terrain. Field elevations range from 2010 feet down to sea level. The cane was transported down the ridge by flume and then by rail in the lowlands. So flat were these miles of coastal lands that there were no brakes on the cane cars. By 1947, trucks had replaced the flumes and railroads.

Kekaha Sugar is the only plantation with a majority of its land leased from the state. It recorded 14 tons per harvested acre in 1983—one of the highest yields of any Hawaii plantation—thanks to the Kekaha and Kokee ditches. In 1994, Amfac/JMB consolidated many functions of Kekaha Sugar and Lihue Plantation as a cost-cutting measure.

7. Oahu

WAIAHOLE WATER COMPANY AND OAHU SUGAR COMPANY

Oahu Sugar Company was established in 1897 on the fertile but dry Ewa Plains in the lee of the Koolau Range. While the lowlands were irrigated from the Pearl Harbor aquifer, pumping water 550 feet up to the *mauka* lands was costly, so the company looked for a mountain source. The nearest source of surface water was on the windward side of the Koolaus. In 1905, Oahu Sugar Company hired engineer Jorgen Jorgensen to explore the possibility of bringing that windward water to Ewa. After several surveys and feasibility studies, the Waiahole Ditch plan—recommended by engineer J. B. Lippencott, assisted by W. A. Wall, and based on Jorgensen's report—was adopted by the directors on 19 August 1911. The Waiahole Water Company was informally established in 1912 to make preliminary arrangements for financing, water rights, easements, and access routes. In 1913, it was formally organized under the agency of Hackfeld & Company with J. F. Hackfeld as president.

The Waiahole Ditch was ambitious by any standard. The initial cost was $2.3 million—and the replacement cost has been estimated to be over $56 million. It started at 790 foot elevation in Kahana Valley, traversed the back of Waikane and Waiahole valleys, pierced the Koolau Range, and ended at the foot of the Waianae Range at an elevation of 600 feet. The original length of the system from Kahana Valley to the terminal reservoir in Honouliuli was 21.9 miles, later extended westward another 5 miles.[1] There were thirty-seven diversions on windward streams. The Waiahole Ditch consisted almost entirely of tunnels. Besides the main tunnel, thirty-eight other tunnels were constructed: twenty-five (later twenty-seven) connecting tunnels on the windward side and thirteen on the leeward side. The shortest was 280 feet, the longest

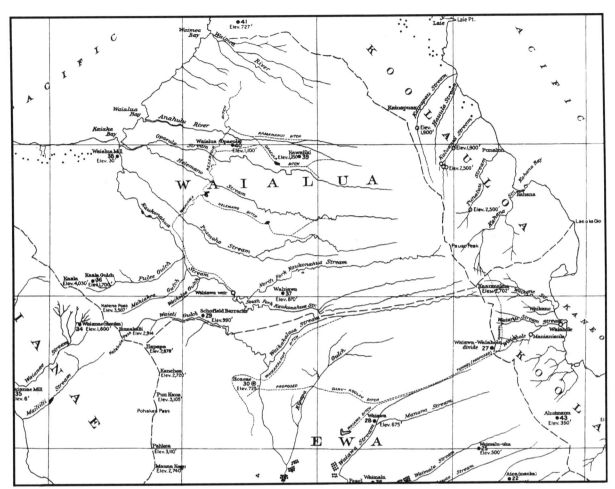

1913 drainage map, Central Oahu. (USGS.)

was 3329 feet. Each one took anywhere from thirty days to a year to bore. The ditches were mostly cement-lined; the reservoirs, dirt-packed.

The main tunnel, which traversed the Koolau Range, was 14,339 feet, or 2.7 miles—the longest transmountain tunnel in Hawaii until the construction of the Molokai Tunnel. The main tunnel's north portal was at an elevation of 754 feet and the south portal at 724 feet. Because it passed under the crest of the mountain, where the dikes were the most frequent, it was by far the hardest tunnel to bore. It was approximately 7 by 7 feet in the arched sections. The grade in the tunnel was 2 feet per 1000; in the rest of the system it was closer to 1.3 feet per 1000.

During the construction of the Waiahole Tunnel, candles relieved the utter darkness: the white wax still drips from niches. (Photo: D. Franzen.)

The windward tunnels and the main tunnel diverted high-level groundwater into the ditch system and conveyed it through the Koolaus to leeward Oahu. After leaving the tunnel on the leeward side, water traveled by pipe and inverted siphon to Kunia and Ewa. There were 1.38 miles of siphon, the longest being 2,034 feet. Subsequently, while the main transmission system remained unchanged, most of the stream diversions were abandoned and several development tunnels added.

The contract was awarded to Hubert K. Bishop, who had previously been territorial supervisor of public works. He estimated the cost at $1.6 million. Before construction began on the tunnel, access roads and trails, camps, storage buildings, and other facilities had to be built. Five to six saddle horses were purchased for the engineers, superintendents, and foremen, along with twelve mules for hauling supplies. A wharf at Waikane was started, and plans were made for a railway from the wharf to the north portal, as well as one from Pump 6 to the south portal.

"Wharf at Waikane." (Photo: R. Heath. Courtesy K. Darling.)

Bishop set about getting his workforce lined up. Engineers W. A. Wall and E. P. Pierce were hired away from Hawaii's Department of Public Works. Men were recruited from the recently completed Hilo Railroad Company. Bishop noted that "we are endeavoring as far as possible to use Japanese and natives for the position of chainmen and rodmen" and, further, "we have endeavored as far as possible to get men of good, clean character and men thoroughly experienced." Skilled labor was recruited from California, New York, Utah, and Pennsylvania. "In addition," Bishop reported, "six experienced shift bosses arrived by steamer from San Francisco today for work in the Main Tunnel. These men are fresh from the Los Angeles Aqueduct, and are said to be the best shift bosses from that work." Bishop reported in February that "the organization is complete and nearly all the men have been selected. In many cases we have not secured the men we most desired on account of the salaries wanted by these men. It has been our endeavor [not to] run the cost for engineering, supervision and administration above a reasonable percentage of the cost of the work."[2]

The most ambitious tunnel project yet undertaken in the Territory of Hawaii was started in February 1913. It was a rocky start. The power supply had not yet been developed, nor had the requisite machinery arrived. The majority of the workforce, initially numbering 600, was located on the windward side. Lacking power and equipment, therefore, the tunnel was started by hand. Incentives of $25 a day were offered to the superintendents for every day the tunnel borings were completed on schedule. Bishop, however, suggested that greater incentives would be necessary.

Problems developed immediately—most of them adding up to extra costs. Although by March and April there were between 800 and 1000 men on the job site, the locomotives were still on the mainland, not to arrive until June. Worse yet, there was still no power for the tunneling, though powerplants were reported to be under construction. The rock encountered at the north portal was so hard that tunneling proceeded at only 2.5 feet a day, working three eight-hour shifts, and was still being done by hand. Moreover, the initial surveying was inadequate. And since the slightest error in tunnel direction would mean that the tunnel faces would not meet, it was necessary to order new surveys constantly, checking and double-checking alignment. Labor costs were questioned by the Waiahole board of directors. Labor was paid the "regular rate" of $1.25 a day—but most of the work was contracted, not by the day, but by the foot. Men in supervisory capacity, the board complained, were getting from $100 to $200 a month. Bishop pointed out that the Department of Public Works was paying their engineers from $200 to $250 a month.

Bishop noted that the laborers were "very careless with powder, and we have already had three serious accidents from the use of powder. In each case the Shift Boss was absent." Things went from bad to worse. In June 1913, typhoid fever broke out in the camps on the Waiawa side, sparking inspection and mandates from the Board of Health. Expenses for a "first aid and sanitary man combined" were collected from the laborers.

Bishop resigned on 1 October 1913. He reported that 912 feet of the Main Tunnel had been driven on the north side, 2063 feet on the south side. Bishop then left Hawaii. Among his subsequent projects he helped engineer the Alaska Highway and became deputy commissioner for the federal Public Roads Administration. In 1945, he returned to Hawaii to assess the war's damage on roads. He died that same year.

Bishop was succeeded by Jorgen Jorgensen, who contracted to complete the tunnels and lined ditches within thirty months. By the time Jorgensen took over work on the Waiahole Tunnel, work had come to a halt. Bridges and fills needed repair. All the tunnels on the north side were encountering hard rock. Storms and rains had demolished the railroad and trails. The locomotive was

"Shay Locomotive at Waiahole, attached to a load of pipe." (Courtesy O.S. Co.)

wrecked with ten men injured. In December, Jorgensen reported that "a series of accidents resulting in four deaths of japanese laborers were accountered (*sic*) during the past month, all owing to carelessness on the part of the unfortunate in handling explosives. They were all men of at least 10 years experience in this kind of work and had been repeatedly warned to be cautious."[3]

The biggest problems were the ones that had already overwhelmed Bishop at the north portal. Besides encountering particularly hard rock, the amount of dike water was unexpected and unparalleled. At 200 feet from the north portal, a dike was penetrated—expelling 2 mgd into the tunnel. At 900 feet the water flow was 26 mgd. The temperature of the water in the tunnel was approximately 66° F. Jorgensen reported: "At the beginning of the tunnel work, three shifts of eight hours each were kept going. This was continued until the large amount of water coming into the tunnel, at North heading, became troublesome, and on account of the hardship on the men, working for eight hours in the cold water, it became necessary to cut the shifts down to six hours each, so that four shifts per day were employed for this heading."[4] Since this was the entrance portal, sloping downward, the two siphons could not keep up with the flow, which submerged the face and made work impossible. Then a second tunnel—16 feet above and almost parallel to the main tunnel—was dug with a slope back toward the north portal. A centrifugal pump with a 12-mgd capacity lifted the excess water to this drainage tunnel.

By June 1914, some 43 mgd was being dumped out of the north portal. This water ran through a 1400-foot penstock and hydroelectric plant that created the electricity for the machinery and lights for both the north and south portal work. Jorgensen reported: "It developed 360 horsepower, converted by means of a 250 kw dynamo into 2300 volts of electrical energy." Jorgensen's contract required that he extend the north face to 1400 feet—a requirement he had fulfilled. He then began to concentrate his efforts on the easier south heading.

Tunneling at the south portal went smoothly at a rate of about 630 feet a month. In September 1914 the tunnel was driven 655 feet—a Hawaiian and, indeed, American record. (The previous record was 649 feet a month for the Elizabeth Tunnel of the great Los Angeles Aqueduct, which was driven through much softer limestone.) The workforce numbered up to 900 workers, mostly Japanese. The tunnel men were skilled professionals who "apparently take delight in the hardships incident to the work, the exposure to the cold water, and the risk in handling explosives. They were on the job all the time and never failed to deliver the goods in situations in which white men or Native Hawaiians would have been physically impossible . . . a bonus being given [the subcontractors] which sharpened their interest and never failed to give results."[5]

When the first dike was breached at 10,530 feet from the south portal, releasing 17 mgd, the railroad track in the tunnel was raised a foot, allowing about 18 mgd to flow underneath. By April the flow was so great that it was difficult pushing upstream with supplies and tools. In June 1915, the face was advanced only 60 feet—and that was considered a prodigious accomplishment. At last the delivery tunnels were made ready, and the dike water was delivered to Oahu Sugar Company in July 1915, well ahead of expectations. Work was then resumed at the north face:

> *For the last three hundred feet the tunnel was a whirling rainstorm, a giant shower bath, a waterspout and a typhoon combined. From sides, roof and face the water spurted in continuous streams, in some instances with a force against which one could not stand. As the airdrills bit into the lavas new streams from the drill holes were frequently such that one man could not drive the steel bars in against them, it required two or three husky miners to force the drills into the holes and hold them there until the air was applied.*
>
> *Into these drill holes the dynamite could not be tamped in the ordinary way. The*

common dynamite stick was torn up and spat out by the flood, so a novel method of loading evolved. Tin cylinders, made to hold ten sticks of giant powder each were prepared, the dynamite unwrapped and jammed into the tins, fused and capped. Then the cylinders were driven against the water into the drill holes and there wedged. It was an uncanny method of using dynamite, but, like a lot of other things being done in the Waiahole work, the only possible way.

Never in mining history has the attempt ever been made to tunnel through such a river bed, the tremendous difficulties of handling everything necessary in and out of a tunnel well over two miles deep, every foot of which is a swirling torrent, being appreciable only to one who has waded that last hundred yards of the main tunnel.[6]

In January 1916, Jorgensen reported that "the two headings of this tunnel met on December 13. Alignment and floor grade were perfect, the length, however, being 124 feet more than the original calculation."

On 27 May 1916, the system officially opened. Due to the difficulties encountered at the north portal, 80 percent of the tunnel (11,700 feet) was driven from the south portal. By September 1916, the final cleanup of the north portal was nearing completion. Camps, tools, and other paraphernalia were being removed. Equipment and supplies were sold to Waianae Plantation, Maui Agricultural Company, and Honolua Ranch. Waiahole Water Company kept fifteen buildings, including three stables, a warehouse and blacksmith shop, washhouse, outhouse, and nine ditchmen's houses. Work continued for the next year on intakes, development tunnels, roads, a pump, and a weir at Waiahole gulch. Charles H. Kluegel was inspecting engineer during the latter stages of the project. Hjalmar Olstad, resident engineer at the time of completion in 1916, remained on as superintendent of Waiahole Water Company until 1948.

The main Waiahole Tunnel has remained unchanged since 1917. Within the pristine atmosphere peculiar to deep caves and tunnels there are still numerous reminders of the men who built it. The wooden pegs in the ceiling, used to carry a line that aided alignment between the surveyor's visits, are still in place.

Although the capacity of the Waiahole Tunnel is approximately 125 to 150 mgd, the transmission system itself can carry only 100 mgd. The system was expected to deliver close to 125 mgd, but in fact this was never achieved. The average daily flow over a 67-year period is 32.2 mgd, which takes into account the originally high flows of stored dike water. As these reserves were depleted and the averages dropped, development tunnels were built on the windward side, starting in 1925. Of the six development tunnels on the north side, four were productive. The greatest of these was the Uwau tunnel. Water pours out of the roof and walls of that tunnel, reminiscent of the original con-

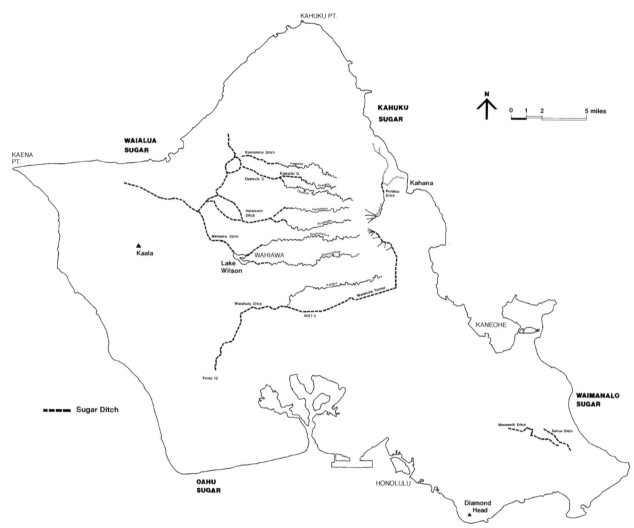

Major sugar plantations and ditches, Oahu.

ditions in the main tunnel. Almost all the water for the Waiahole Ditch is developed by tunnel—it is calculated that surface water diversion contributes only about 2 mgd of the average daily flow of 28 million gallons. This flow supplied Oahu Sugar Company, which used on the average of 140 mgd. The balance came from groundwater.

In 1970, Oahu Sugar Company acquired Ewa Plantation. The 1890 Ewa Plantation Company annual report noted that the plantation commenced with

Workmen at Waiahole Tunnel adit. Tunnel engineering was a highly technical business, and few projects offered more challenges than the Waiahole Tunnel. (Photo: R. Heath. Courtesy K. Darling.)

"5 white men, two natives, 8 Chinamen, 4 horses and 8 mules." This statement belies the major investment required to launch a viable sugar plantation. A more realistic approach was reported in 1895:

> *The enterprising projector of the O.R. & L.Co., B. F. Dillingham, is planning to utilize a portion of the Ewa land in the establishment of a new sugar plantation to eclipse in magnitude all that have preceded it in these islands. There are some 10,000 acres of choice lands all of which may be irrigated by a system of wells with pumping plant. The concern will incorporate with a capital of $2,000,000 in shares of $100 each, much of which is already subscribed for.*[7]

By 1931, Ewa Plantation had seventy artesian and four surface wells with eighteen pumps. Its total pumping capacity in the summer of 1931 was 118 mgd—one and a half times the amount used by the entire city of Boston. This large and successful plantation is little discussed here because it did not develop any water from surface sources. But its groundwater use, combined with that of

Table 5
Waiahole Ditch Development Tunnels

Tunnel	Date Built	1990 average (mgd)
Kahana Tunnel	1929–1931	3.6
Waikane Tunnel	1925–1927, 1934	4.5
Waikane Tunnel 2	1927–1929	1
Uwau Tunnel	1932–1935*	13
Tunnel B		Caved in
Tunnel A		Abandoned

*Extended in 1963.

Oahu Sugar Company and Honolulu Plantation, amounted to over 200 mgd from the Pearl Harbor aquifer to irrigate sugar on the Ewa Plains. The comparatively small amount of water from Waiahole supplemented this, a reminder of sugar's endless thirst.

Waiahole Water Company changed its name to Waiahole Irrigation Company (WIC) in 1970. Today it is a wholly owned subsidiary of Amfac/ JMB–Hawaii. Water from the Waiahole Ditch was used by Oahu Sugar Company until its close in 1994. Within a year, however, the state and the people were embroiled in a bitter hearing before the Commission on Water Resource Management to determine the future allocation of the Waiahole water. Indeed, the issues are seen as the benchmark for the future of water allocation, conservation, and stream restoration in the entire island chain.

WAIALUA SUGAR COMPANY

Sugar cultivation and milling in Waialua can be traced back to at least 1865, with the Levi and Warren Chamberlain Sugar Company. That company failed, as did several that followed, until, in 1889, Waialua Agricultural Company, known later as Waialua Sugar Company, was incorporated by Castle & Cooke. The earlier failures were due in large part to insufficient capital investment. Castle & Cooke, however, was determined not to repeat this mistake, so within the first years of ownership it proceeded to expand the acreage, build a new mill, put in a railway system, and develop an adequate water supply, drawing from both ground and surface sources. An immediate measure of the success of this strategy was the increase in sugar production from less than 5000 tons in 1900 to nearly 20,000 tons in 1905.

Waialua Sugar Company is distinguished by the efficiency of its storage and irrigation system. The distribution of water is especially flexible: the ditches are so interconnected that nearly all the water can be sent to any given place. This plantation has four surface-water collection systems—the Wahiawa, Helemano, Opaeula, and Kamananui—all built between 1900 and 1906. The Wahiawa–Lake Wilson system, by far the largest, delivers 10 to 12 billion gallons a year, Helemano around 700 million, Opaeula 350 million, and Kamananui 90 million. In short, Waialua Sugar Company has the largest water storage capacity in Hawaii.

Although the company's annual report of 1902 reported the Helemano and Poamoho ditches under construction, there is some confusion regarding this date, as another reference indicates that the Helemano lands were not acquired until 1905. The Helemano system diverted Helemano and Poamoho streams. The *mauka* section of the Helemano Ditch is the Tanada Ditch. Although designed to handle large flows, this ditch system averages only around 3 mgd.

The Opaeula Ditch was completed in 1903 at a reported cost of $12,000. It tapped the three major tributaries of the Anahulu River: Kawainui, Opaeula, and Kawaiiki. The next year the Kawainui was tapped at a higher elevation by the Kamananui Ditch. This ditch was redesigned for a 30-mgd capacity in the mid-1920s, and at the same time the Opaeula Ditch was realigned so that it became independent from the Kamananui system. The Ito Ditch, built sometime after 1911, took water to the Mokuleia area.

The key to Waialua's irrigation was Wahiawa Dam and Reservoir. The 2.5-billion-gallon capacity reservoir was completed in two years, on 23 January 1906. It was the largest reservoir in Hawaii and the most economical as well. Later known as Lake Wilson, it provided 90 percent of Waialua Sugar Company's surface water. At a cost of only $300,000, it was a brilliant investment. H. Clay Kellogg, a hydraulic engineer from Santa Ana, California, was chief engineer for both the dam and the ditches, assisted by Eugene Valgean of Anaheim, California.

The dam itself, at 136 feet, is the highest earthen dam in Hawaii. Sited at the 1000 foot elevation, it measures 461 feet long and is 580 feet thick at the base. It created a 7-mile-long reservoir that took advantage of the natural streambeds and canyons located in the Kaukonahua gulch. The source was 8000 acres of watershed at the head of the Koolau Mountains. Lake Wilson was fed by a ditch system known first as the Oahu Ditch and later as the Mauka Ditch Tunnel. It consisted of 4 miles of main ditch and 8 miles of laterals, which included thirty-eight tunnels. It was started in June 1900 and completed in March 1902 at a cost of $80,000. The capacity of this ditch system was 90.5

mgd. Besides developing water in the Kaukonahua watershed, it also diverted from the Poamoho watershed.

Another 4 miles of ditch, tunnel, and siphons delivered the water from Lake Wilson (as well as from Helemano and Opaeula ditches) to Waialua's upper fields at 730 feet elevation. This Wahiawa Ditch had a capacity of 50 mgd. The total cost was $49,177.59, making it one of the least costly projects of its size, averaging out to $1.59 a lineal foot. Of the ditch's 20,740 feet, only 1600 feet was in open ditch. The remaining length comprised twenty tunnels, the longest of which was 1742 feet. It had the largest and tallest flume on Oahu: 130 feet high. In 1923, most of the flumes spanning the gulches were replaced by siphons.[1]

By 1924 Waialua Sugar Company had thirteen pumping stations—three steam-powered and the rest electric. By 1931 it was ranked fourth in the islands in terms of sugar production. In 1951, the remaining pumps were electrified to improve the efficiency of the system along with major changes in the reservoir, pipelines, and distribution ditches. In 1957, Waialua Sugar opened Pump 17, a 150-foot vertical shaft with two 1500-horsepower pumps. This well averaged 15 mgd and was used to irrigate the midlevel fields. Lake Wilson stored storm water and serves as a flood control, and in more recent times received effluent from the Wahiawa sewage treatment plant. Almost all of its 10,000 plus acres was irrigated using an average of 90 mgd. In a typical year, two-thirds of this flow was provided by groundwater, the balance by surface water.

Getting the water was one challenge; putting it on the fields efficiently was yet another. Waialua Sugar Company became known for the success of the Waialua Flume, used for field irrigation. This system of concrete flumes was designed in portable sections that were placed in the fields in a herringbone configuration; water was then released into the furrows by small tin gates. Using these flumes the plantation boasted one of the highest man/day performances per acre irrigated ever recorded in the industry. In 1995, it was the last surviving sugar plantation on Oahu. Despite the successes of Waialua Sugar Company, Castle & Cooke announced its closing in 1995. The fate of Lake Wilson, which the city had come to depend on for the disposal of effluent from its sewage treatment plant, became yet another concern for Hawaii state and local government.

KAHUKU PLANTATION COMPANY

Kahuku Plantation was started in 1890 by James Castle and Alexander Young on land leased from James Campbell. That same year the manager reported that dry weather kept the pumps "agoing night and day most of the time," tying up labor and adding to operational expenses by using a substantially larger amount of coal over the previous year. In 1902, Alexander & Baldwin became the agent for Kahuku. The plantation was handicapped by its small size (about 2500 acres in cane) and its difficult land, soil, and climate. Water was supplied entirely by groundwater sources, which yielded around 45 mgd. In 1906, Kahuku Plantation Company expanded to Punaluu lands with a lease from the Bishop Estate giving it access to more lands and to Punaluu stream.

The Punaluu Ditch system consisted of two ditches. The larger was the 1.8-mile Punaluu Ditch, to the north of the stream, with an average flow of 10 mgd; mostly concrete-lined, it included twelve short tunnels. The second ditch was unlined and extended less than a mile to the south of the stream, hooking around toward Kahana.

In 1931, Kahuku Plantation increased its fields by leasing Laie Plantation agricultural lands from Zions Securities for twenty-five years, although it had been milling Laie cane prior to this time. Kahuku Plantation became a wholly owned subsidiary of A&B in 1968. It closed in 1971.

WAIMANALO SUGAR COMPANY

The Waimanalo Sugar Company was established in 1878. It was one of the smallest plantations in Hawaii, averaging less than 10,000 tons per year. Most of its water came from the Kailua watershed.

Kailua Ditch, the earliest of Waimanalo Sugar's three ditches, diverted water from upper Kailua springs in the Waimanalo basin and emptied into the Waimanalo Reservoir. A second ditch, built in 1924, had its source in the Kawainui Swamp. Two pumps lifted the water from that swamp and took it to the head of a 10,000-foot system of small tunnels, mostly through stone or hard earth, into a reservoir. This ditch cost $220,000.

The ditch most associated with the Waimanalo Sugar Company is the Maunawili Ditch. Its source is high-level tunnels, springs, and streams in Maunawili and Waimanalo Valley. The dirt- and cement-lined ditch includes about twenty flumes, many measuring no more than a foot and a half each way, before it crosses through the Olomana Tunnel to Waimanalo. During dry seasons this ditch delivered less than 2 mgd. Waimanalo Sugar eventually had 99 percent of its sugar under irrigation—and nearly 25 percent of that came from surface water sources.

Waimanalo Sugar Company closed its sugar operations in 1947 and surrendered its water license in 1953, which then reverted to the Territory of Hawaii. Today Hawaii's Department of Agriculture maintains the Maunawili

While most ditches were wise and sure investments, a few were economic black holes of repairs, such as Hawaii's Upper Hamakua Ditch and Oahu's Waimanalo system, shown here. Both systems eventually reverted to the Hawaii government, which spent a great deal of money on them with, some would argue, no better results. (Photo: D. Franzen.)

Ditch, which serves about thirty farmers in Waimanalo with an estimated minimum flow of only 54 million gallons a month. Visually, this was a gem of a system up until recent times. Although small, it had all the components of a typical ditch system: flumes, ditches, tunnel. Its particular charm was its redwood flumes, which were in remarkably good condition in 1984. These have since been abandoned in favor of PVC pipe.

8. East Maui

EAST MAUI IRRIGATION COMPANY

The alphabet soup of Hawaii's companies gets especially thick on Maui. Samuel Alexander and Henry Baldwin were the founders of Alexander & Baldwin (A&B) and East Maui Irrigation Company (EMI). These two men started their illustrious career together in an informal partnership in 1869 with the purchase of 11.94 acres of Bush Ranch. In 1876 they formed the Hamakua Ditch Company and in 1878 completed the Hamakua Ditch—not to be confused with the 1904 Hamakua Ditch Company on Hawaii, which later changed its name to Hawaiian Irrigation Company, or that company's Upper and Lower Hamakua ditches.

During the ensuing decade Alexander and Baldwin's plantation was incorporated as the Paia Plantation and included Haliimaile Plantation (Grove Ranch), East Maui Plantation, and Seaside Farm. The agency of Alexander & Baldwin was established in 1894. The corporate partners gained control of Hawaiian Commercial and Sugar Company (HC&S) in October 1898, and Alexander & Baldwin then became agent for HC&S. It was a meteoric rise for the two men—from the new firm of Alexander & Baldwin, which had posted a net profit of $2627.20 in 1895, to A&B, Ltd., which had accumulated assets of $1.5 million at the time of its incorporation in 1900.

Immediately on acquiring HC&S, the partners started the Lowrie Ditch —also known as the Lowrie Canal—which started in the rain forest of Kailua in Makawao district. The ditch had two sources. The first was a reservoir at Papaaea that was fed by two five- to six-mile ditches. The second source was Kailua stream where the diversion intercepted the source of the older Haiku Ditch and ran parallel to that ditch. (The old Haiku Ditch was abandoned between 1912 and 1929.)

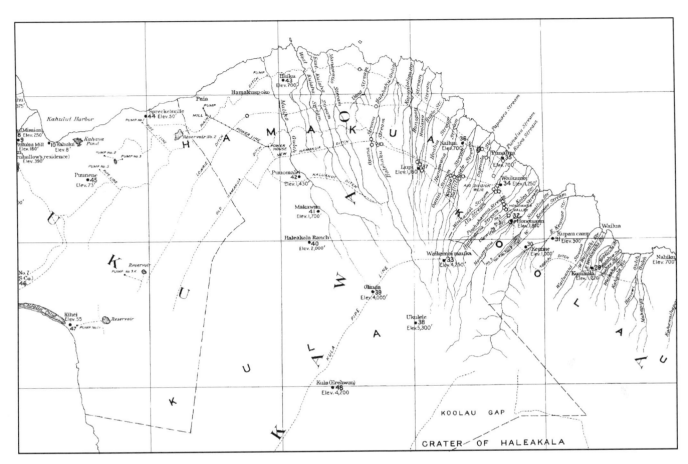

1913 drainage map, East Maui. (USGS.)

The ditch was named after William J. Lowrie, manager of HC&S's plantation and mills at Spreckelsville. It was designed by engineer E. L. VanDerNeillen and supervised by Carl Jensen, who was reported in 1900 to be on his way to his "old home" in Denmark to recuperate. The work was done by Japanese laborers "under the supervision of one of the brightest Japanese in the Islands." Contracts were signed in July 1899; the work was finished in September 1900; the cost was $271,141. With a capacity of 60 mgd, it was capable of irrigating 6000 acres. This 22-mile system was three-quarters open ditch and included these elements: seventy-four tunnels for a total of 20,850 feet, the longest being 1955 feet; nineteen flumes for a total length of 1965 feet; and twelve siphons with a total length of 4760 feet, the biggest being 250 feet deep

Ditch trails such as this one in the back of Honomanu Valley, Maui, reached into many pristine Hawaiian valleys to access ditches and tunnels. (Photo: D. Franzen.)

at Halehaku gulch. This ditch, by means of inverted siphons, ended at the 475 foot elevation, 257 feet above the Haiku Ditch.[1]

The next big project for the Hamakua Ditch Company was the Koolau Ditch, built in 1904–1905 under engineer M. M. O'Shaughnessy. The Koolau Ditch extended the water collection system another 10 miles toward Hana, around the Koolau Range to Makapipi, in 1904. The cost of Koolau Ditch was $511,330. Its capacity was 85 mgd. This ditch traveled through more difficult terrain than most other systems, and it presented greater logistical problems. O'Shaughnessy reported:

> *The country was so steep and precipitous that little ditching could be employed, and it was necessary to make four and one-half miles of wagon road and eighteen miles of stone paved pack trails to facilitate during construction the transportation of supplies. About 4000 barrels of cement and 100,000 pounds of giant powder were used. In all ten mountain streams are intercepted, which are admitted into the main aqueduct through screens of grizzly bars spaced three-quarters of an inch apart.*[2]

There were 7.5 miles of tunnel and 2.5 miles of open ditch and flume. The thirty-eight tunnels, all dug out of solid rock, were 8 feet wide and 7 feet high. In length they averaged 1000 feet: the shortest was 300 feet and the longest 2710 feet. A total of 4.5 miles of 6-inch-thick concrete lining was used in the tunnels:

> *The work was all done by Japanese with hand drills; ore cars were employed in moving the excavated materials, and it has cost finished about $7 per lineal foot. The Japanese make excellent miners and rock men, and, owing to their small size, it was practicable to work four in a face, and, by working three 8-hour shifts, the whole work had to be completed in 18 months from the date of commencement, April, 1903.*[3]

The Koolau Ditch was later turned over to EMI, who lined and improved it at a cost of $385,117. Originally it fed into the New Hamakua Ditch at Alo, but it was connected to the Wailoa Ditch upon its completion in 1923.

On 23 June 1908, Alexander & Baldwin formed the East Maui Irrigation Company to succeed the 1876 Hamakua Ditch Company. Its purpose was to develop and administer the surface water for all the plantations owned, controlled, or managed by Alexander & Baldwin. The EMI boundaries were from Nahiku to Maliko gulch and included all the area where surface water was developed. West of Maliko gulch was HC&S. In that same year, A&B gained control of Kihei Plantation.

Ditch building continued apace under the newly formed company. The New Haiku Ditch was completed in 1914 with a capacity of 100 mgd. It was mostly tunnel, partially lined, with a length of 54,044 feet. Kauhikoa Ditch was completed in 1915 with a capacity of 110 mgd and a length of 29,910 feet. Wailoa Ditch was started in 1918 and finished in 1923. Mostly tunnel, all lined, with a length of 51,256 feet, it had an original capacity of 160 mgd, later increased to 195 mgd. Once the ditch systems were completed, EMI then turned to building water development tunnels.

EMI's collection system had 388 separate intakes, 24 miles of ditch, 50 miles of tunnels, and twelve inverted siphons as well as numerous small

The Wailoa Canal has a greater median flow than any river in Hawaii. Water collected here at the Wailoa forebay drops through a low-head 500-kilowatt hydroelectric power-plant. (Photo: D. Franzen.)

feeders, dams, intakes, pipes, and flumes. Supporting infrastructure included 62 miles of private roads and 15 miles of telephone lines. The water source was primarily surface runoff from a total watershed area of 56,000 acres. Of this watershed, EMI owned 18,000 acres—the 38,000-acre balance belonged to the State of Hawaii. The state issued four licenses, named Huelo, Honomanu, Keanae, and Nahiku, to EMI for water arising on government land. Each license was initiated at a different time and dealt with differing conditions. The value of the water was determined by its accessibility and distance from fields, and the price was tied to the price of sugar. The state's share was determined by the percentage of rain falling on government land.

Haiku Ditch, Maui. Water-monitoring stations are on many ditches and streams. The gauging program in Hawaii is a cooperative effort between the USGS and the Hawaii state government. Historically the primary purpose of the monitoring program has been to monitor the flow for sugar diversions. This purpose, however, must be revised to meet the needs of the present. (Photo: D. Franzen.)

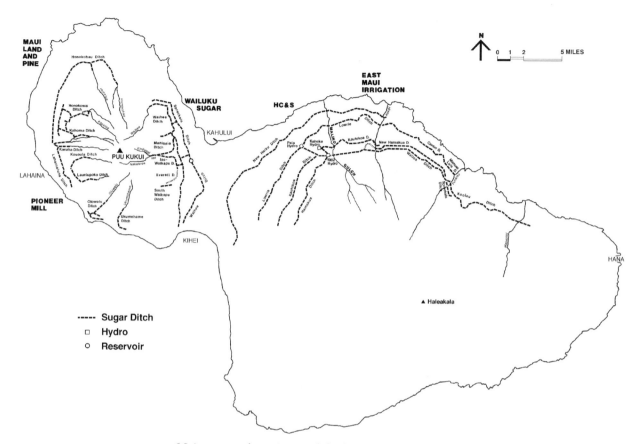

Major sugar plantations and ditches, Maui.

The huge and complex EMI system has developed and changed over the years at a cost of nearly $5 million. The replacement cost is estimated to be at least $200 million. Among the water entities, none compares to EMI. It is the largest privately owned water company in the United States, perhaps in the world. The total delivery capacity is 445 mgd. The average daily water delivery under median weather conditions is 160 mgd, although this ranges from 10 to 445 mgd. Its largest ditch, the Wailoa Canal, has a greater median flow (170 mgd) than any river in Hawaii. EMI supplies Maui County between 850 million and 1 billion gallons of water per year for domestic purposes.

East Maui Irrigation controlled only surface water to HC&S—groundwater was controlled by HC&S itself. But EMI could not always supply enough water to meet plantation requirements, which ranged as high as 200 mgd. Thus, as in many of Hawaii's plantations, groundwater was a major source of supplemental irrigation water.

By 1931, HC&S was able to pump 144 mgd. To accomplish this it relied on deep and powerful pumping stations. Station 2 had equipment at 119 feet; Station 3, called Kihei A&B, had an underground chamber at 300 feet. The deepest was pump 18 at 500 feet. In dry times, pumps supplied up to 45 percent of the irrigation water. Pump 7, which struck water at approximately 125 feet, had a capacity of 40 mgd, and in 1931 was the most powerful pump in the world. It is one of several designated as "Maui-type basal water tunnel," which used a skimming tunnel to collect fresh water off the top of the basal lens. HC&S also received West Maui water from the Waihee Canal and Spreckels Ditch through agreements with Wailuku Sugar Company. By 1931, HC&S was producing about 32 percent of Hawaii's total sugar crop.

Maui Agricultural Company was formed in 1921 by the merging of seven small East Maui plantations: Haiku Sugar Company, Paia Plantation, Kailua Plantation, Kula Plantation, Makawao Plantation, Pulehu Plantation, and Kalialinui Plantation. HC&S, based in Puunene, and Maui Agricultural Company, based in Paia, merged in 1948, at which time Alexander & Baldwin owned about 35 percent of the stock of each company. This merger consolidated all of A&B's sugar plantations on Maui under HC&S. In 1962, HC&S merged with and became a division of Alexander & Baldwin, and EMI became a subsidiary of A&B.

EMI currently has four parallel levels of water development ditches, running from east to west across the East Maui mountains. From *mauka* to *makai* these are the Wailoa, New Hamakua, Lowrie, and New Haiku ditches. The Lowrie runs at a considerably lower elevation than the Wailoa, taking advantage of groundwater development between the two. Wailoa and Lowrie run all the time; New Hamakua and New Haiku run on surplus water from the other ditches or for delivery to the fields. Little remains of the early Hamakua and Haiku (Spreckels) ditches.

The last of the four state-issued water licenses to EMI expired in 1986. A&B and EMI alternately hold revocable year-to-year permits from the State of Hawaii at flat monthly rates.

9. West Maui

WAILUKU SUGAR COMPANY

Wailuku Sugar Company was first organized in 1862 by James Robinson & Company, Thomas Cummins, J. Fuller, and agent C. Brewer & Company, which gained controlling interest two years later. It was incorporated in 1875. Wailuku Sugar Company took over Waihee Plantation in 1895, at which time Spreckels' 1882 Waihee Ditch became the source of conflict and legal action. Wailuku Sugar protested that Spreckels did not have a proper right-of-way across what was now its land. It further disputed his right to Waihee stream water and took the matter to court. But Spreckels lost control of HC&S long before this issue was resolved.

The new owners of HC&S shared a common interest with Wailuku Sugar Company in a proposal for a second ditch to divert the Waihee stream at a higher elevation. Since the parties (Wailuku Sugar and HC&S) were still in court over issues involving the Waihee Ditch, they negotiated an interim exchange lease agreement in 1904. The terms of the agreement—made permanent with exchanges of fee title almost twenty-five years later—were that HC&S got five-twelfths of the new Waihee Canal water and one-half of the older Waihee (Spreckels) Ditch water; maintenance cost for these ditches was shared in the same proportion. Further, HC&S got all surplus water from all ditches and 100 percent of the water from the South Waiehu Ditch. The Happy Valley development tunnel was shared by both plantations, but HC&S got first draw. Waihee water went to HC&S by way of the Spreckels Ditch from 7 PM to 5 AM each night. Also, HC&S relinquished 9693 acres of land in Waikapu, Maalaea, and Wailuku to Wailuku Sugar Company. With these issues resolved, Wailuku Sugar undertook the Waihee Canal.

These terms were not particularly unusual. In many ways they were based

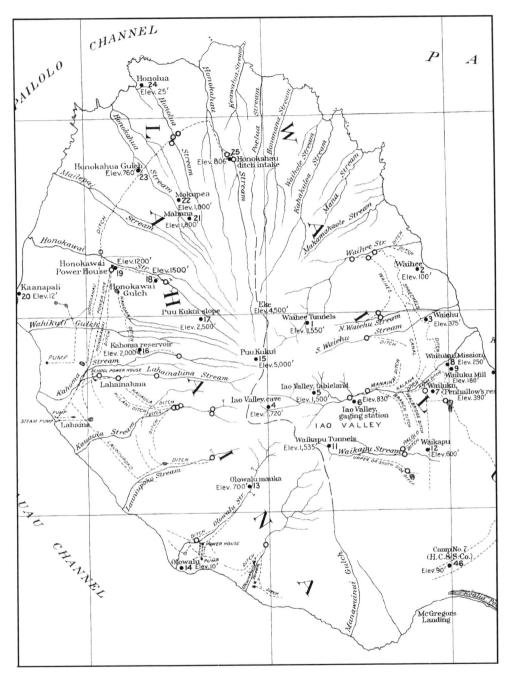

1913 drainage map, West Maui. (USGS.)

on Hawaiian allocation practices as they were understood at the time. It was traditional Hawaiian practice to divide use by time of day and by source. It was also part of the Hawaiian tradition that maintenance was proportionate to use. But when these agreements were made between the sugar companies and small private landowners, it was the small landowners who seemed to get the night hours and second draws.

The Waihee Canal (Waihee Ditch) was started in June 1905 and completed in May 1907. The cost was $160,000. Built under the direction of engineer James T. Taylor, this 50-mgd-capacity ditch tapped the Waihee stream at the 650 foot elevation, just below the Aliele Falls. This intake, however, was abandoned in favor of a new one slightly downstream in 1935. Its 10.62 miles included twenty-two tunnels totaling 16,539 feet; thirty-nine flumes totaling 2764 feet; 35,549 feet of open, cement-lined ditch; and a 1253-foot-long, 3-foot-diameter siphon to cross Iao Valley. Ditch grade averaged 2.5 feet per 1000. The longest tunnel (2246 feet) was especially challenging, as much of it went through hard close-grained rock and it required compressed air and percussion drills. This tunnel took eighteen months to cut. The contract price for the labor ranged from 85 cents to $5 per foot, depending on the material cut, the location, and the length of the tunnel.[1]

James Townsend Taylor was one of several engineers who worked on a number of projects throughout the islands during the ditch building decades. Before coming to Hawaii, Taylor had worked for the Southern Pacific Railroad in charge of tunneling and construction. He arrived in Hawaii in 1898 and worked for Honolulu Sugar Company, Kekaha, Waialua, Wahiawa, Kona, Haiku, Paia plantations, and the Kahului Railroad Company. He was consulting engineer for the Wailuku and Kahului waterworks.

Wailuku Sugar Company ditch names, it must be noted, are particularly confusing. In recent times the newer ditch (formerly the Waihee Canal) has been referred to as the Waihee Ditch and the older one as the Spreckels Ditch (formerly the Waihee Ditch). But there is a Spreckels Ditch on East Maui, as well, which was originally known as the Haiku Ditch.

By 1913, Wailuku Sugar Company was irrigated entirely from mountain sources. Besides the major ditches discussed here, it had nine smaller ones: two on Waiehu stream, five on Wailuku stream in Iao Valley (the largest was the Maniania Ditch), and two on Waikapu stream (South Side and Palolo ditches). Some of these have since been abandoned or consolidated. Wailuku Sugar and Maui County cooperated in water development tunnels in the Waikapu and Wailuku valleys. The Maniania and the Iao–Waikapu ditches shared the

Water is collected in this gravel trap before being sent on its way in the Iao-Maniania Ditch. (Photo: D. Franzen.)

Wailuku stream intake. The Waikapu intake of the Everett Ditch was replaced in 1933, and that ditch was eventually abandoned when a landslide filled the intake and tunnel. The South Waikapu Ditch in the Waikapu Valley irrigated the field directly below. The North Waiehu Ditch delivered water to the Waihee Ditch. The Kama Ditch intake was below the Iao–Waikapu intake, in the Wailuku stream. This ditch mostly served *kuleana* lands.

Wailuku Sugar Company ended sugar production in 1988. Since then its water has been used primarily for cultivation of macadamia nut orchards and pineapple crops.

HONOLUA RANCH AND PIONEER MILL COMPANY

In Hawaiian, "Lahaina" means cruel sun, and indeed the place was known for its heat and droughts. This side of West Maui was perfect for growing sugarcane because it was fertile and sunny, but water was scarce. Most of the streams were small—perennial only in their headlands. The largest water sources on this side of the island were the Honokohau and Honolua streams, on land to the southwest owned by Henry P. Baldwin known as Honolua Ranch. Honolua Ranch operations included selling livestock, wood, coffee, watermelons, onions, corn, alfalfa, potatoes, and pineapple, as well as running a store. But it did not raise sugarcane. That was left to Pioneer Mill Company, with lands on the Lahaina side of Honolua.

Pioneer Mill Company traces its history back to 1860, when James Campbell, Henry Turton, and James Dunbar started the Lahaina mill. The cane was supplied by independent sugar planters. The company was incorporated in 1885 and acquired by H. Hackfeld and Company, the predecessor of Amfac (now Amfac/JMB). Pioneer Mill obtained water by reaching back into the valleys with eight separate collection systems: Honokohau Ditch (built and owned by Honolua Ranch) plus seven smaller systems it built itself. By 1931, Pioneer Mill received from 50 to 60 mgd from these sources and an additional 40 mgd was supplied by pumping groundwater.

The success of Pioneer Mill depended on satisfactory water agreements with Honolua Ranch, which controlled the major water-producing watershed. (Honolua Ranch changed its name to Baldwin Packers, then Maui Pineapple Company, and finally, in 1969, to Maui Land and Pineapple Company, commonly referred to as Maui Land & Pine, or ML&P.) The two companies agreed that Honolua Ranch would build and own the Honokohau Ditch, while Pioneer Mill would finance it and use the water. Honolua would repay those costs with interest at 6 percent in semiannual installments of $5000. Pioneer Mill stipulated a minimum capacity of 20 mgd. Once the ditch was completed, Pioneer Mill would buy up to 15 mgd for the first five years at $3000 per million gallons a day per annum and thereafter at the reduced rate of $2750. There would be no charge for water in excess of 15 mgd. Honolua Ranch would maintain the ditch at its own expense.

The Honokohau Ditch was completely built twice and renovated once, all in the span of twenty years. The survey for the first Honokohau Ditch was done by J. S. Malony in November 1901. He designed a system that was primarily ditch and flume with five inverted siphons crossing the gulches. He estimated the cost at $132,000 or $120,000, depending on which elevation was used. The total length was to be 53,240 feet, or 12.5 miles,[1] of which only 16,300 feet would be tunnel. The plan called for tunnels measuring 6.5 feet deep and wide.

The Honokohau Ditch was begun in 1902 and completed in June 1904. The ditch started at an elevation of 700 feet, with intakes at Honokohau, Kalua-nui, and Honolua streams. The capacity was variously reported as 39 mgd and 30 mgd. The ditch and flumes ran along soft hillsides where landslides frequently choked the ditch and damaged the flumes. Less than two years later Malony, noting the deteriorating condition of weirs and flumes, urged that the system be better maintained. He stressed that to wait on repairs until deterioration was advanced would be much more costly than a good ongoing program. Nevertheless, David T. Fleming, manager of Honolua Ranch, reported in 1912 that "the old Ditch continues to leak worse every day."

Fleming determined to build an entirely new ditch to replace the 1904 system. The new ditch was to be called the Honolua Ditch to avoid confusion with the existing Honokohau. (It is now referred to as the Honokohau Ditch by Pioneer Mill Company and Honolua Ditch by Maui Land & Pineapple.) The new alignment ran roughly parallel to the older ditch and was based on a study and survey by Jorgen Jorgensen. Fleming was distressed at Jorgensen's "far too high" bill, which was $2706.62, and thought his cost estimates were inflated, too, though he allowed that the report "probably is a good job." Fleming originally reported that "labor appears to be plentiful, and there are men asking for ditch work almost every day." But that situation soon changed: "We are very short of labor for the cementing, as our tunnel gangs invariably head for Oahu as soon as they are paid off in full; and it has been necessary to import new men to do the cementing. The one cement gang now at work is taking hold well, and we believe we will be able to locate more experienced men soon, probably on Hawaii."[2]

Work commenced in June 1912. John Harrison Foss, the engineer in charge, was from Stanford University, where he taught in the engineering department: he came to Maui during the summers and supervised the work. Ultimately, though, Hawaii called to Foss, who moved to Maui in 1919 and managed East Maui Irrigation Company for the next eighteen years. Two features minimized seepage in this new ditch: not only was it totally lined, but wooden flumes were eliminated, relying on tunnel instead. Indeed, the final plan called for thirty-one contiguous tunnels—the longest was tunnel 17 at 4740 feet—separated by adits. Altogether there was 34,241 feet of tunnel, 726.3 feet of covered crossings, 1183 feet of inverted siphons, and only 427.3 feet of open ditch.

Mules were used to do the hauling over tracks laid in the tunnels and up

Ditch Lining. Next to tunnel work, cement work was the most important, costly, and time-consuming part of most ditch projects. Lining protected against water loss, bank erosion, and tunnel collapse. Plantations were constantly on the lookout for economical and effective ditch lining to reduce the considerable loss from seepage into the porous ground. The most successful lining was 4- to 6-inch cement, but this was expensive. Jorgen Jorgensen and others experimented on the Hamakua Ditch with plaster reinforced with chicken wire. While this eliminated the form work required in cement-lined ditches, plastered ditches tended to break up. Maui Agricultural Company and HC&S experimented with the use of a cement gun, but reports indicate there was no advantage to the device. In 1917, Wailuku Sugar was lining ditches with precast concrete slabs, allowing the water to keep flowing during the work. Many plantations used cut fieldstones set in cement mortar to line the ditches. While the work was time consuming and highly specialized, the impervious stone did stand up well.

Honolua Ditch: "Hand Mixing Concrete for tunnel bottom." (Photo: D. Fleming. Courtesy ML&P.)

Honolua Ditch: "Small mixer at work showing 'merry go round' charger which practically doubled its capacity." (Photo: D. Fleming. Courtesy ML&P.)

Honolua Ditch: "Putting in crossing #26–27 with small mixer." (Photo: D. Fleming. Courtesy ML&P.)

Honolua Ditch: "Big mixer at work— loading into cars which ran directly into the tunnels." (Photo: D. Fleming. Courtesy ML&P.)

the trails. A rock crusher was brought on the project to provide gravel for the concrete. The rock encountered during the tunneling was on the whole not excessively hard, but in tunnel 17 some hard rock was encountered. Except for tunnels 17 and 19, all tunnels were driven by hand. Jorgensen's plan called for plaster lining; the floor lining, however, was upgraded to 2- to 3-inch-thick cement. Although Fleming noted that the plastered sides appeared to be standing up fairly well on the Hamakua Ditch on Hawaii, in a summary at the completion of the Honolua project he said: "Our only regret is that we did not make an absolutely permanent job by concreting the sides as well as the bottom."

The ditch was completed by November 1913. It was with considerable pride that Fleming could report that "an unusual feature of the construction was that there was not a single fatal accident." The ditch cost $239,841. Its capacity was 50 mgd. The new ditch was a great success, setting a record of 791.42 million gallons delivered in February 1914. Honolua Ranch did not retain any water for its various agricultural ventures—all water went to Pioneer Mill. Fleming commented: "The formerly incessant cry of the Pioneer Mill Company for more water appears to be quieted for the present, and an ominous silence reigns."

The increased water delivery was not the only measure of success, for by November 1914 total revenues from water sales ($65,378) comprised over half of Honolua Ranch's revenues for that year to date. The Honolua Ditch was clearly a good investment.

Despite the care taken to seal the ditch, by 1921 a seepage study showed a 31 percent loss from intake to weir. This loss was unacceptable. For a five-year period starting in September 1923, Pioneer Mill undertook to reline the entire Honokohau Ditch— at its own expense, it appears. It purchased dump cars and two Ford locomotives from Maui Agricultural Company. When the dump cars proved too wide for the tunnel, undaunted they dismantled and rebuilt them to fit. Pack mules were used to haul the cut stones for the ditch lining, each animal carrying four stones in boxes built especially for that purpose.

To ensure that the flow of water to Pioneer Mill was not interrupted during the repair period, the water was diverted into the previously abandoned 1904 Honokohau Ditch, a section at a time. Because the old ditch had to be renovated for this purpose, the workers were actually working on both ditches. And because they could only work a section at a time, they could not work on all tunnels simultaneously as they had during the construction phase. Thus while it took just eighteen months to build the ditch, it took five years to reline it. At the same time, the ditch capacity was increased from 50 mgd to 70 mgd.

Honolua Ditch: "Machine Drilling #17." G. N. Wilcox's first tunnel was dug with hand tools by two Chinese workers in two weeks. With the advent of explosives, tunneling became quite a different thing. As described by William Moragne for the 1898 Huleia Tunnel: "You would dig a coyote hole—that is, a tunnel into the bank, then a cross tunnel on the end. Put in about a ton of powder, plug the hole and blow her up. . . . Or, there was another way. Use a churn drill to go down about 20 feet into the bank. Set off a small explosive at the bottom of the hole to hollow out a bigger space. Then lower 2 or 3 kegs of black powder (giant powder) and blow it up." While this process got more sophisticated over the years, it remained in essence the same. (Photo: D. Fleming. Courtesy ML&P.)

Once the lining was completed, the work gangs dismantled the old Honokohau Ditch flumes and filled in much of the old ditches. Some 1904 stone-lined ditches remain, as do the tunnels, now used to gain access to the intake. By the third time around, this ditch was one of the most expensive ones in cost per foot.

In addition to getting water from the Honokohau Ditch, Pioneer Mill had seven smaller water collection systems of its own. The largest of these, and best documented, was the Honokowai Ditch. The earliest effort to collect water from Honokowai stream was by way of a semicircular galvanized iron flume built in 1898. But by 1917 these flumes had so deteriorated that the company decided to replace them with a tunnel, increasing capacity tenfold and providing some degree of permanency. Because this necessitated crossing government lands, in 1917 a twenty-one-year license to cross these lands was negotiated. Detailed progress reports of the building of the Honokowai Ditch give us an ongoing description of the many aspects typical of these large projects.

The timing of this project coincided nicely with the completion of Waiahole irrigation system. Jorgen Jorgensen offered to sell 3000 feet of 16-inch blow pipe for 25 cents a foot and about a ton of drill steel left over from the Waiahole project. Not only did Pioneer Mill purchase equipment from the Waiahole Water Company, but it gained engineer F. Koelling from that project as well, when their original engineer, Mr. Young, was called up by the National Guard. The consulting engineer was R. Renton Hind.

The camp was set up—housing, walls, and roof were made of corrugated iron—equipment and supplies were obtained, and work began. As usual, all the workforce, including Koelling, lived at the site in camps. A house for Koelling was established at the camp, and it is presumably for this purpose that the following items were ordered from the company's Puukolii Store in February 1917:

> *One Bed, 1 Spring, 1 Mattress, 4 Stove Pipes, 1 Stove Cap, 1 Stove Roof Plate, 1 Fry Pan, 1 Sauce Pan, 1 Kettle, Spoons, 1 Bowl, 1 Lamp, 1 Strainer, 1 Kitchen Knife, 1 Dish Pan, 1 Tea Pot, 1 Can Opener, 1 1/2" Hose Bibb, 1 Pr. Cup & Saucer, 1 1/2" Union, 1 1/4 Yd Screen Wire, 1 Pr. Drawer Pullers, 1 Pr. Butts.*[3]

The logistics and mechanics of this project were documented in a fairly exhaustive report. For example:

> *The machinery for the north heading consists of one 10 × 10 Laidlaw-Dunn-Gordon compressor with 50 HP motor, one size 5 type "B" American blower with 15 HP motor, one Allis-Chalmers motor generator set of 18 1/2 K.W., one 3 1/2' × 10 1/2' air receiver, one three ton Jeffrey locomotive with Edison battery, cars, rails, pipe, lumber, transformers, &c, and was hauled by Holt caterpillar traction engines to an elevation of about 2,500 feet, directly above the north portal.*
>
> *At this point, an "A" frame with sheaves, deadmen, &c., was erected and a 7/8 steel cable run to the opposite side of the gulch and anchored to a number of large trees, about 300 feet above the bottom of the gulch and opposite the portal. The cable was drawn taught (sic) by the 45 HP caterpillar engine traveling down the mountain road, after which the load was hung on two sheaves running on the cable and slacked down by a three inch manila rope until it was directly above the bottom of the gulch, when both cable and rope were slacked together and the load landed at the portal. The heaviest load weighed 2¼ ton.*[4]

At the south portal the machinery was

hauled by Holt caterpillar engines directly to the portal and consists of one 16 × 14 and 10 × 14 Duplex, Imperial type 10, Ingersoll Rand compressor with 100 HP motor direct connected, one 12 1/2 K. W. Allis-Chalmers motor generator set, one 4 × 6 × 8 duplex Worthington pump, and one Jeffrey electric locomotive, transformers, &c.

A blacksmith shop situated at the main camp about half way between the two portals is equipped with one 8 × 8 Laidlaw-Dunn-Gordon compressor, 20 HP motor, #635 Leyner sharpener, forge, drill press, emery wheels, &c. The steel is sharpened with a four and five point regulation Leyner die and plunged for tempering.

In the north heading, two No. 18 Leyner drills and four sets of 1 1/4" hollow steel from 24" to 96" are used, one machine drilling six feet per hour and averaging 1,150 cub. ft. of free air at 80 lbs. pressure per lineal foot of hole drilled. The water for the drills and compressor is taken from the stream at an elevation far enough from the portal to furnish 45 lbs. pressure. The air line is two inch, and the water line one inch galvanized iron pipe.

In the south heading two No. 9 Leyner drills and four sets of 1 3/8 inch, hollow steel from 24" to 96" are used. The pressure maintained at the compressor is 90 lbs. per square inch, and there is a large surplus of air. The water for the drills is forced into the tunnel and compressor by a four inch Worthington duplex pump, regulated to about 40 lbs. pressure and operated by air.

Three shifts of two drillers, two helpers, three muckers and a shift boss are worked, and a seven ft. round of from 10 to 16 holes is drilled in seven hours, one hour being taken up in setting of bar twice, loading holes, &c.[5]

The concrete-lined tunnel, approximately 6 by 6 feet, started at an elevation of 1525 feet. For blasting through the primarily soft rock, "40% gelatine (*sic*) powder, triple tape fuse, and No. 6 caps are used, fifty pounds of powder breaking about 4 1/2 feet."[6] Because ventilation was a serious problem during construction of the main tunnel, due to the excessive slope, additional blowers were placed in both faces—after each blasting it took them at least twenty minutes to clear the air so the men could return. The muck was removed in side-dump Koppel cars of 1-yard capacity and hauled by a 3-ton Jeffrey electric locomotive. Besides candles, 50 and 75 watt, 110-volt Mazda lamps were used to light the tunnel while work was going on.

Nakamura, the contractor for the main tunnel, asked for $3.50 a foot plus the usual plantation bonus. Koelling suggested that he get $4.00 for the north face and $3.50 a foot for the south, but no bonus. This variation in rates, typical of contract tunnel work, was determined not only by economics but by the degree of difficulty encountered. It was not unusual for rates to change during the project as difficult conditions were met, for the men simply would not work below a certain wage. The Amalu development tunnels were contracted to Iwasaki for $2500, which worked out to $3.00 a foot. Wages for a mucker were $1.50 a day. Drill helpers got $1.75 a day, drillers $2.00 a day, shift bosses $2.75 a day. Bonuses, often substantial, were given on top of this. During one month, for instance, the contractor received an additional bonus of one-third the net amount earned of $1230. If progress exceeded 400 feet in a month, there was an additional 25 cents a day per man.

Water flowed through the 1.55-mile Honokowai tunnel in March 1918. Built for a capacity of 50 mgd, the average of the Honokowai system was 6.15 mgd. Water sources were the Amalu and Kapaloa branches of the Honokowai stream; later, an additional source was Honokowai Well 1.

Pioneer Mill lost the right-of-way for the first section of the main tunnel in the late 1940s. When the plantation was forced to abandon that section, it built a second tunnel on the adjoining land, which it owned. The work was done by Pioneer Mill's civil engineering department. That tunnel was completed in 1950 at a cost estimated at $151,342. The next year the plantation acquired the land that the original section traversed. Today there are two tunnel mouths sitting 30 feet from each other at the Honokowai intake.

Pioneer Mill's other six ditch systems were smaller yet: each one irrigated the upper fields directly within each watershed. They were named the Kahoma, Kanaha, Kauaula, Launiupoko, Olowalu, and Ukumehame, after the streams they tapped or areas they served. In addition to delivering water to the fields, the Kauaula Ditch powered a hydroelectric plant, Olowalu Ditch supplied domestic water to Olowalu village, and the Kanaha Ditch supplied the Lahainaluna Ditch, Maui County, and the Pioneer Mill factory.

Of these six systems, the upgrade to the Kauaula Ditch is the best documented. The original Kauaula Ditch was replaced by a tunnel in 1929. As a fairly complete report survives, we are afforded a glimpse into the day-to-day operations of ditch and tunnel building. Work was started in April 1928 with the surveying and building of the camp. The lumber for the four camp buildings came from the old Honokohau camp. As in most of these projects, pack mules were used to transport materials. It was not until late September that the camp was complete, however, with enough area cleared for the hoist, compressor, crusher, and warehouse buildings. The Yamamoto Gang then began excavation. The monthly report states: "The tunnel workmen are exhibiting an excellent spirit and the inexperienced ones are eager to add to their knowledge of this type of work as shown by the monthly progress. The total number of men on the Kauaula Tunnel work is 30, 11 of them being Filipinos and 19 Japanese."[7]

Concrete tunnels and ditches were the most durable and water-efficient. (Courtesy Bishop Museum.)

Pioneer Mill experimented with air-driven jackhammers for the first time, which it reported were economical and fast. Kohler "electric light machines" were also considered very satisfactory. Pioneer Mill reported: "The new No. 1 S. Berger & Sons transit, which was ordered for the establishing of the tunnel alignment, has been tried out and proved to be an excellent instrument with the exception of the compass needle. This latter is somewhat sluggish but can be remedied at later date. Such a transit makes accurate work possible."[8]

Tunneling proceeded around the clock on the Kauaula Tunnel. An accident was reported on the evening shift of 4 October 1929:

One of the drill operators, Kawamoto, and Hashimura, the night foreman, were engaged in drilling holes in the approach to the tunnel with a Jackhammer. Just before the holes were completed an unexploded charge of dynamite was struck with the Jackhammer and an explosion followed. Kawamoto was killed instantly while Hashimura died on the way to the hospital. Both were skilled tunnel men but Hashimura was one of the most promising young men in our employ.[9]

Misfortunes continued. Soon after that explosion, the surveyor took a bad fall off a cliff and the engineer developed serious leg problems. These accidents were given additional significance by the men on the job, for a burial site had been unearthed. Then, as now, finding a grave was considered a bad omen on a construction site, and this event required considerable attention to moving the remains and having everything properly blessed. The bones were reinterred and work resumed.

When it was completed, the Kauaula Tunnel had a total length of 4013 feet including adits (106 feet). It was cement-lined, with a slope of 20 inches to 1000 feet, and had a carrying capacity of 25.5 mgd and a median of 4.5 mgd. Water was used to run Kauaula hydroelectric plant before being sent to the fields.

In the search for water, Pioneer Mill dug numerous development tunnels but met with little success. There was one short-lived exception when a tunnel at around the 2600 foot elevation succeeded in developing 6 mgd. Reported flowing in 1904 by M. M. O'Shaughnessy and in 1908 by George B. Sturgeon, this would have been the most successful water development tunnel in Hawaii at the time. It may well be that this tunnel had tapped and released a finite supply of high-level water. It is not even mentioned in Stearns' 1942 survey of Maui's water systems, however, which reported twenty-two high-level tunnels on West Maui, none of which were very productive.

Of the remaining five ditches, only bare statistics survive. The Kahoma Ditch started at an elevation of 1930 feet and had a second intake at 960 feet. Water from the upper intake traveled through a mile-long tunnel to the fields; that from the lower intake went through a long wooden flume approximately 2 by 2 feet. There were three known water development tunnels above Kahoma —one dry, one that developed about 0.01 mgd, and one that developed about 2 mgd. The median flow of the Kahoma Ditch is 3.19 mgd.[10]

The Kanaha Ditch had two intakes; the combined median was 3.68 mgd. The water from the upper intake ran through an 8-inch pipe. Maui County had rights to the first 0.5 mgd and the State of Hawaii had rights to four and a half hours a day of the total flow for use at Lahainaluna School (though in practice the school uses what it needs). Pioneer Mill sold surplus water to the county when its own needs for the factory were satisfied. Overflow went into Lahainaluna Ditch.

The median flow of the Launiupoko Ditch was 0.78 mgd. There is one known development tunnel above the intake that was reported to develop 0.1 mgd.

In the early 1930s, Olowalu Plantation merged with Pioneer Mill Company, bringing along its two small and relatively crude systems, the Olowalu and Ukumehame ditches. The Olowalu system had a capacity of 11 mgd and a median of 4.08 mgd. The Ukumehame had a capacity of 15.5 mgd and a median flow of 3.30 mgd.

By the 1980s, Pioneer Mill no longer received all of the Honokohau Ditch water: 6 mgd, more or less, was allocated elsewhere. Honokohau Valley farmers got about 1 mgd, Maui Land & Pine used some for its pineapple fields and the Kapalua resort, and Maui County took several million gallons daily for domestic use. From 1961 to 1981 the average delivered to Pioneer Mill was 24 mgd. During the drought of 1983–1984, the flows were the lowest in its history and the ditch was often dry by the time it reached the plantation. The flow in this ditch has been closer to 18.6 mgd in recent years.

Pioneer Mill was one of the last five surviving plantations in 1996. Just as Lahaina's hot, dry climate was perfect for sugar, given enough water, so it was well suited to tourism, another high-water-consumptive industry. At the same time, however, the Honokohau farmers were petitioning for more water for their use and for the stream itself. It looks like competition for water had met the limits of the resource in Lahaina.

10. Hawaii

KOHALA DITCH COMPANY

In 1900 there were five plantations, each with its own mill, in Kohala. Small in size and poor in water, each plantation struggled independently to develop its water supplies. Kohala Sugar Company experimented with water development tunnels but with little success. At Niulii, a promising tunnel built in 1899 dried up in a few years. Halawa Plantation had three new reservoirs of "questionable suitability" due to an irregular supply from the streams. E. E. Oldings, manager of Hawi Sugar Company, irrigated about 600 acres with groundwater, but it was hard going. He got "one mgd at about 85 ft elev. Supply brought to the surface by a coal-fed Worthington compound engine and then to an elevation of 200 and 460 feet by Worthington triple expansion low duty engine . . . a very expensive one to operate."[1] The second tunnel delivered about 2 mgd.

Oldings maintained that irrigation was worth even these costs. In fact, he attested that the yield from one field increased from 4 to 9 tons per acre as a result of irrigation—a remarkable increase as well as a most healthy yield for the period. It might be that Oldings exaggerated in order to emphasize what all sugar planters know: irrigation is essential.

Not only was the water costly, it was easily contaminated. At the Kohala Sugar Company's pumping station, water salinity went from 10 to 25 grains of salt per gallon in five months. While this concentration was well within the acceptable range, the rapid change suggested that there was a limited supply of fresh groundwater.

It was clear that these individual attempts at water development by small independent plantations were inadequate: survival would depend on consolidated effort. The trustees of the Bishop Museum and Bishop Estate, having ownership of adjacent watershed lands, commissioned engineer Arthur S. Tuttle

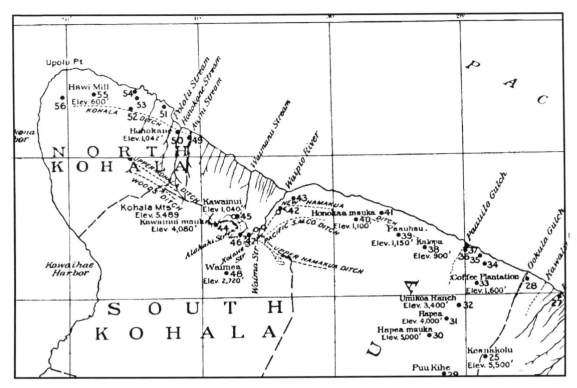

1913 drainage map, Kohala-Hamakua. (USGS.)

of Brooklyn, New York, to again evaluate the possibility of bringing water out of the Kohala–Hamakua watershed. The Tuttle Report was completed in 1902. In Tuttle's opinion, the project was feasible.

As a result of the Tuttle Report's findings, two ditch companies were formed. The Kohala Ditch Company, established in 1904, built the Kohala Ditch. The Hamakua Ditch Company, formed around 1906, built the Upper and Lower Hamakua ditches (not to be confused with the Hamakua Ditch Company, started on Maui by Alexander & Baldwin, or with their 1878 Hamakua Ditch).

On 12 March 1904, J. S. Low acquired a license from the Territory of Hawaii for a period of fifty years to "enter upon, confine, conserve, collect, impound and divert all the running natural surface waters on the Kohala-Hamakua Watershed," a license that he assigned to the Kohala Ditch Company (KDC). The license stipulated that Low service all requesting customers.

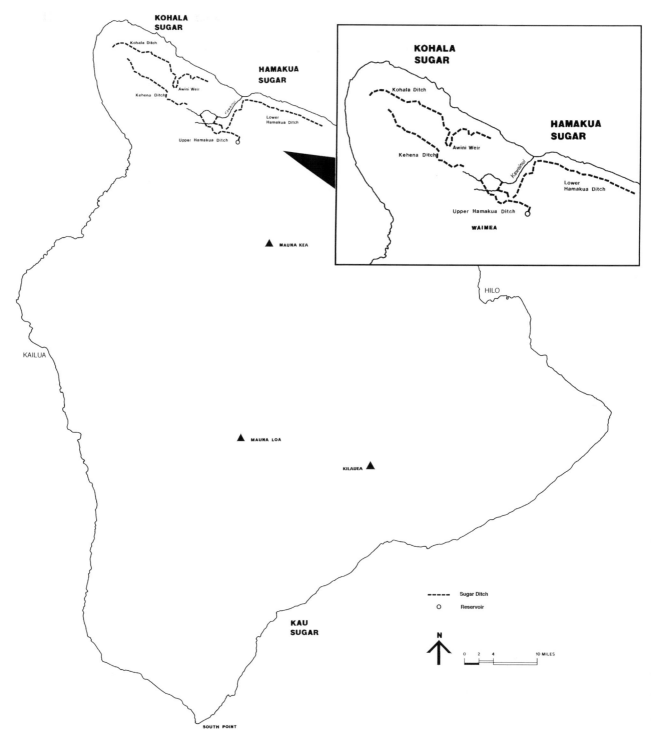

Major sugar plantations and ditches, Hawaii.

Shortly after, leases were acquired from the Bishop Estate, owner of much of the watershed. The new waterway would be called the Kohala Ditch.

The Kohala Ditch Company approached the small plantations for commitments to buy the water. It proposed a choice: either a large ditch, with a capacity of 70 mgd and a water fee of $3000/mgd per annum, or a small ditch with a capacity of 16 mgd and a water fee of $2500/mgd per annum.[2] The decision was to go with the larger ditch. Union Mill, Kohala Sugar, Niulii Plantation, Hawi Sugar, and Halawa Plantation all agreed to participate in the project, committing their companies to the purchase of a certain amount of water. This minimum ranged from 2 mgd to 6.75 mgd in the first year, but most of the plantations contracted for more in ensuing years. A few remained skeptical. Even in 1906, after the ditch was completed, Robert Hall of Niulii complained about the cost of the water. He closed his letter by saying: "I trust that irrigation may prove a success with us, but I have my doubts."[3]

The Kohala Ditch Company was financed through the sale of stocks and bonds, most of which were bought by Charles Hart, who also had interest in Union Mill, Hawi Sugar, and Niulii Plantation. Another KDC stockholder and one of the "three most enthusiastic" backers of the project was John T. McCrosson; two others were Michael O'Shaughnessy and Samuel Parker. These last three names would resurface on the Hamakua coast and in the Hawaiian Irrigation Company.

The Kohala Ditch contract was signed on 1 March 1905 with the requirement that the ditch be completed in fifteen months. O'Shaughnessy was the chief engineer; Jorgensen was his assistant. P.W.P. Bluett was the engineer in charge. The Kohala Ditch turned out to be the best-documented and most dramatic of O'Shaughnessy's projects. Without holding any records for the earliest, largest, longest, or hardest to build, the Kohala Ditch was nevertheless one of the most ambitious ditch projects in Hawaii. Perhaps no other was constructed over such difficult terrain. Because the water traveled in fairly straight courses through mountain tunnels, construction of the access trail, cut on the edge of precipitous cliffs, was a more fearsome task than building the ditch itself.

The entire ditch was about 23 miles long. According to O'Shaughnessy: "About half the work was inaccessible to wagon roads, and narrow pack trails about five feet wide were cut out of the sides of the mountain precipices and used for packing material."[4] Like most of the ditch systems, its mountain section was composed mostly of tunnels. There were fifty-seven tunnels, in fact, the longest being 2500 feet. Over the total length there were about 16 miles of tunnels and 6 miles of open ditch, the remainder being flumes. The ditch was lined primarily with stone or cement as far as Hawi; the extension beyond

Hawi was unlined. The tunnels were built with tapered sides to hold the cement lining. The flumes were "of the most substantial kind, thoroughly calked and tarred, being 7 feet wide by 6 feet deep." There were 600 Japanese contract laborers. A contemporary report said:

> *The Japanese are indifferent to weather conditions, rain and exposure seeming to keep them in a healthy condition, while white men, similarly exposed, would be disabled by rheumatism and other ailments . . . . These men seemed to like this class of work and made a success of breaking rock where other nationalities were a failure. As their wages average only $1.00 per day for the ordinary classes of lava rock encountered, hand drilling with Japanese is more economic than the use of either air or electric drills.* [5]

A comment by O'Shaughnessy suggested that the laborers had some reservations about "this class of work" at a dollar a day: "While prone to listen to the advice of agitators, compared with similar bodies of white men working under similar conditions their conduct may be considered conservative." [6]

A hospital and medical department were provided for the men "who were assessed 50 cents /mo. apiece for this object." O'Shaughnessy continued:

> *All the men were housed under corrugated iron roofs, framed by light scantlings, such as 2×3 inches, which could be easily packed. The side walls were made of eight-ounce duck tacked to the scantlings and the floors were 1×12-inch, nearly always two feet above the ground and higher if practicable, so as to provide a place for drying the men's clothes in wet weather. Such accommodation was found much better and healthier than tents, which quickly rotted under the heavy rains.* [7]

A 1907 report, however, contradicted O'Shaughnessy's rosy view of his workers' capacities:

> *. . .The boring of the tunnels was an impressive and dangerous feat, the openings in some places being almost inaccessible until trails were cut out of the mountain sides. Six men and many mules were killed by falling down these precipices in the progress of the work. Most of the boring was through soft rock, though machine drills at times were required. The sturdiest of Japanese workmen would succumb after a few weeks labor in the darkness and cold, and emerge, emanciated (sic), for a season of hospital treatment.* [8]

The Kohala Ditch was built in two sections: the Honokane and the Awini. The Honokane section was opened on 11 June 1906. At the opening ceremonies Mrs. Samuel Parker set the waters of Honokane flowing to the Kohala, Niulii, Halawa, Kohala, Hawi, and Union mills with these words: "I christen thee Kohala Ditch. May you bring blessings, happiness, and prosperity to the people of Kohala." According to a contemporary report:

Waiahole Ditch: "Packing Timbers *mauka* on the North Side." (Courtesy O.S. Co.)

Mules on the Ditch Trail. Pack mules were the backbone of many ditch building projects, hauling men and materials to and from the site. On the Honokohau Ditch, pack mules were used to haul the cut stones for the ditch lining, each animal carrying four stones in boxes built especially for that purpose.

Honolua Ditch: "Track was laid the length of the ditch, and all supplies as well as sand, cement, etc., were hauled by mules." (Photo: D. Fleming. Courtesy ML&P.)

The opening of the ditch was made an occasion of considerable rejoicing and the event passed off with much éclat. Beside the interested and prominent people of the district, quite a party of Honoluluans attended by special trip of the Kinau to Mahukona for the occasion, at which speech making and prophetic utterances in keeping with the importance of the undertaking were indulged in by several of the party. The Acting Governor in his address said: "It marked a new industrial era for our people, and was an event that in some ways was the most remarkable that has yet occurred in these islands."[9]

The Awini section was finished in 1907, as was the extension to Puukea Plantation. The Awini section started from the Waikaloa stream and traveled over 8 miles, mostly in tunnel, to the Awini weir. Here the water dropped 900 feet in a manmade waterfall into the Honokane section.

The Kohala Ditch trail stands out as one of Hawaii's most beautiful trails, but it is also one of the most dangerous, beset with landslides and impassable during big rains. As M. L. Randolf concluded sixty years later: "Considering the lack of adequate maps, the rugged terrain and the extreme isolation of the construction sites, this must stand as an engineering and construction feat of very considerable magnitude. The original construction stands for the most part unimpaired and unrevised to this day."[10]

Prior to construction, project backer John McCrosson had promised the plantations delivery of 70 mgd and had in fact quarreled with O'Shaughnessy's estimate of 30 mgd. O'Shaughnessy's conservative prediction, however, was the more accurate: on the average, the Kohala Ditch delivered 22 to 30 mgd. (The higher flow figures recorded in the earlier years included short-term dike discharge.) The minimum flow has been variously reported as 8 mgd and 3.5 mgd. The capacity of the ditch was originally 70 mgd, later reduced to 50 mgd when the original flumes were replaced with smaller ones. Half the flow in the Kohala Ditch was provided by the Awini section. During dry seasons, however, that section did not run, as it was dependent on surface runoff. In 1946 the Honokane tunnel was extended 375 feet, making a total length of 1862 feet. Twenty-three dikes were tapped, yielding an additional 5 mgd.

The ditch changed from a gathering and transmission system in the mountains to a delivery and distribution system at the Niulii weir. However, a little water is added beyond the Niulii weir through the Iole system, which the ditch leased from the Bond Estate. This system depended on surface runoff, and from 1965 to 1970 fluctuated from a low of 1.5 million gallons a month to a high of 192 million gallons a month. High flows usually could not be used, however, since they came in times of plentiful water.

The Kohala Ditch Company's water agreements demonstrate the complications that can arise when several watershed landowners are involved. The

Honopue Bridge, Hawaii. The Kohala Ditch trail is dwarfed by the surrounding mountains. The Sproat family is closely associated with this ditch and trail. Jacob William Sproat came from Missouri and worked on construction of the Kohala Ditch. He succeeded ditch superintendent Lyman Perry, a position that paid $75 a month in 1917. When Jacob retired, his son Bill succeeded him. Bill was in turn succeeded by his son Dale, who held that position until 1977. (Photo: D. Franzen.)

lease rent was based on the water yield from each parcel. But quantifying that yield presented certain difficulties: the amount of water delivered could fluctuate; water monitoring was minimal, unreliable, or nonexistent; and water developed during rainy periods, when it was not necessary to irrigate the fields, was less valuable than stable sources during dry times.

About 40 percent of the Kohala Ditch crossed government lands, although in more recent agreements it was estimated that 50 percent of the water arose on those lands. About 5000 acres of state land naturally drain into Awini Ditch above the 2000-foot level at Apua, Waikapu, Honopue, Awini, and Puukapu. There were about 770 acres of state lands above the 1000-foot level in Pololu. The rest of the ditch was primarily over Bishop Estate lands, which included the Honokane stream source. Due to its consistently reliable flow, Honokane was considered the most valuable of the sources.

The cost of building the Kohala Ditch was low: under $600,000. The Honokane section cost $407,789; the Awini section cost $98,346; the 1908 Wailoa extension cost $46,782. The Kohala Ditch Company's balance book,

Table 6
Kohala Ditch Company Revenues: 1920–1934

Year	Water revenues	Maintenance cost
1920	$63,000	$26,700
1921	$88,500	$24,700
1922	$95,500	$18,400
1923	$97,500	$22,900
1924	$89,500	$21,400
1925	N/A	N/A
1926	$92,100	$19,100
1927	$96,600	$15,800
1928	$104,000	$16,700
1929	$90,900	$18,700
1930	$63,800	N/A
1931	N/A	N/A
1932	$96,200	$10,400
1933	$92,700	$16,600
1934	$85,900	$18,300
Total	**$1,197,250**	**$201,500**

Note: Figures are rounded to the nearest hundred.

1920 to 1934, shows sufficient income from water revenues to cover construction costs as well as maintenance costs plus a profit. Is this a good measure of how other ditch companies fared economically? It is hard to tell. Bookkeeping practices and water fees may well have been very different in those cases where the ditch company was essentially one and the same as the plantation it served—and did not need to demonstrate to its stockholders a profit discrete from plantation activities.

The Kohala Ditch Company also owned much of the commercial and residential land in Kohala and Hawi. Its second main source of revenue was from rental of these lands. In 1934, rent collected from fifty-seven businesses and public buildings, as well as from 150 individuals, totaled $87,400, or slightly more than the water revenue for that year. The Kohala Ditch Company, it seems, was a sound investment.

The Kehena Ditch Company was financed initially by the Kohala Ditch Company, later joined by Hackfeld & Company. It started the Kehena Ditch in 1912 and completed it in 1914. Twenty percent of this ditch crossed state land. It terminated in the Puukumao Reservoir, which in the 1960s was determined to be unsafe and was breached. The ditch had been largely abandoned by that time anyway, as it was never a reliable source of water. The state initiated an effort to redirect the water from the upper section of the ditch to Kawaihae and Hawaiian Home Lands, but this effort was interrupted by new federal clean water standards which would have required additional treatment of the water, making it uneconomical. The ditch was abandoned, although Kahua Ranch takes water by pipe from the upper segment, which still runs.

In the 1930s the Kohala Sugar Company was formed by the consolidation of five smaller plantations under the agency of Castle & Cooke and became the primary user of Kohala Ditch water. The Kohala Ditch Company then became a wholly owned subsidiary of Kohala Sugar Company. In 1975, however, Castle & Cooke closed Kohala Sugar. As there were now no customers for the newly available water, Castle & Cooke reduced maintenance costs. When the flume at the back of Honokane broke, water from the Awini section could no longer reach the Honokane section of the ditch. The Honokane section itself was minimally maintained. As the tunnels filled up with debris and gravel and the flumes began to deteriorate, less water traveled through the ditch. The open ditch beyond Hawi was allowed to fall into disrepair. Most of Kohala Sugar Company's land and assets, including the Kohala Ditch Company, were eventually purchased by Chalon, a Japanese company. Chalon has restored some of the ditch and delivers water as far as Upolo.

HAWAIIAN IRRIGATION COMPANY

Once the Kohala Ditch was completed, the chief participants turned their attention to developing water for the plantations on the Hamakua coast. The Hamakua coast plantations, much like those in Kohala, were numerous, independent, and in competition with each other. Many of them were represented by different agents, who were each to some degree intent on gaining control over the rest. This situation was quite different from almost everywhere else in Hawaii, where one or two plantations dominated large areas, often united under a single agency. The impact of this arrangement on the water companies was considerable: instead of being a subsidiary of a plantation, these companies were beholden to their investors to make a profit. The resulting tension among the various companies and individuals was especially evident on the Hamakua coast.

The Hamakua Ditch Company was formed in 1904 with president and chief stockholder Harry Lewis. Together Michael M. O'Shaughnessy and Jorgen Jorgensen surveyed the Upper and the Lower Hamakua ditches. When O'Shaughnessy left the company after a disagreement with John T. McCrosson in January 1907, Jorgensen became engineer of the company. The Upper Ditch was built between June 1906 and August 1907; the Lower Ditch, from May 1909 to June 1910.

The Hamakua Ditch Company transferred its agency from Lewis & Company to Theo. H. Davies & Company in November 1909 and changed its name to the Hawaiian Irrigation Company that same year.[1] Among the local subscribers to the Hawaiian Irrigation Company (HIC) were F. A. Schaefer & Company, Pacific Sugar Mill, Hamakua Sugar Company, Allen & Robinson, H. Hackfeld, Ahrens, and Jorgensen. Davies underwrote company losses with a $70,000 overdraft with the banking house of Claus Spreckels & Company as well as a $150,000 loan, in essence owning the company and its assets, the Upper and Lower Hamakua ditches.

The plantations on the Hamakua coast did not depend on irrigation to the degree that most other plantations did. They enjoyed greater annual rainfall than almost any other area in Hawaii. Even when used, irrigation was minimally effective when applied on the extremely porous soils of the area. Consequently, water from both the Upper and Lower Hamakua ditches was used primarily at the mills and for fluming, not for irrigation. In 1923, Honokaa Sugar Company's manager recommended overhead irrigation after seeing it at Hawi, but this did not happen until the late 1940s under operational manager Richard Frazier. It was not until the late 1970s or early 1980s, with the introduction of drip irrigation, that irrigation was cost-effective on the Hamakua coast.

The Upper Hamakua Ditch—a hasty job comprised of dirt ditches and galvanized flumes patched with lumber—was opened before it was properly lined. It was completed in January 1907, at a cost of $240,000, and initially delivered 15 mgd. The water sources were the Kawainui, Alakahi, and Koiawe tributaries of the Wailoa stream, which runs through Waipio Valley, plus surface runoff as the ditch cut through the watershed, plus water that was developed in the tunnels. The ditch was hailed with much enthusiasm. Some water must have been applied to irrigation, for in 1909, Hamakua Sugar Company manager K. S. Gjerdrum reported to Schaefer & Company an increase in average yield at his plantation from 2.89 to 4.06 tons per acre and wrote: "And I have come to the conclusion that we do not dream of the effect on the growth of the cane the lower Ditch water will have—it will revolutionize things in general."

By 1915, the average flow of the Upper Ditch had diminished to 8.1 mgd, slightly over half its initial delivery. During dry times, up to 85 percent of the water was lost through seepage by the time it got to the end of the line.[2] The flumes and mountain access trails were badly deteriorated. The Upper Ditch was repaired and upgraded at a cost of $74,139. It boasted some of the best dressed-stone work in Hawaii. The contractor specified that "nothing but redwood, which is good for 15 to 20 years, should be used in such a wet country."[3] Jorgensen noted in 1921 that 65 percent of the Upper Ditch had been rerouted and the remaining 35 percent had been enlarged and improved: "Not a foot of the old line is in use today as originally constructed."[4] In 1925, the ditch was cemented up to the government road at Mud Lane. In 1928, further cement work was halted for financial reasons. In 1935 the Waikoloa section was relocated.

Despite all the work, this was not a successful ditch. It was a black hole for endless repairs from the moment it opened. Although unlined ditches worked elsewhere in Hawaii, in the porous Hamakua soil they proved impossible. Even after it was lined, the ditch was plagued with an inconsistent water source. Ditch flow depended primarily on freshet runoff—in other words, it ran in the rainy season, when water was superfluous, and was dry during periods of drought.

On 1 August 1948, HIC declined to renew the water license for the Upper Ditch and it was taken over by the territorial government. There was no maintenance on this system until the late 1980s, when $5 million of repairs

These stones were hauled by mule for the Upper Hamakua Ditch. (Photo: D. Franzen.)

and reconstruction was done by the state in the Alakahi and Koiawe sections. The section to Puu Alala was abandoned and the water was piped to Hawaiian Homes, Puukapu Homesteads farmers, and Lalamilo Farm lots, all in Waimea. Although nothing of the original 1907 ditch remains, several miles of the 1915 work still survive.

The Lower Hamakua Ditch, "another of the great enterprises which are bringing this island towards the fulfillment of its destiny,"[5] was a happier story for HIC investors. Work commenced in 1909 and the ditch was opened 1 July 1910, delivering water to Pacific Sugar Mill, Honokaa, and Paauhau plantations. It tapped the same water sources as the Upper Ditch, but at lower elevations. There were 21 miles of survey trails and 30 miles of pack trails. The ditch itself had an overall length of 24.75 miles; the cost was $795,214.

With an initial average of 40 mgd, this was a successful ditch. The peak flow was 61.2 mgd—but at this point the facilities started to break apart under

Introduced plants slowly took over the landscape, framing this ditchman's house on the Honokohau Ditch. (Photo: D. Franzen.)

the pressure, so the system was rated at a capacity of 60 mgd, later lowered to 45 mgd. It maintained a fairly consistent flow of 30 mgd. While initial expenses were high and repairs by 1915 amounted to $22,805, the profit in that same year was $185,000.

In the mountains, a series of forty-five tunnels extended 8.9 miles and included "tunnels which are among the longest single borings in irrigation projects in the world"—the longest was 3312 feet. It was further distinguished by its impressive size: approximately 10 by 12 feet. Drills and compressors and gasoline engines were used for tunneling, and telephones were installed along the whole system. The project used 100 pack mules, 200 tons of powder, 25 tons of candles, and 6000 barrels of cement for the tunnel (8000 for the ditches). One million board feet of timber went into flumes, camps, and permanent buildings. As the laborers themselves paid for ammunition and lighting, they tended to be quite economical in the use of these materials.

As in all the ditch projects, labor costs comprised the major proportion of total expenses for the Lower Hamakua Ditch. The total expenditure for labor through June 1909 was $40,000, of which $30,000 was for contract labor. (Materials for the same period were only $22,973.) At that time about 650 men were employed. Later, about 1200 men were on the job,

> *chiefly composed of Japanese, Hawaiians and Koreans, with a fair sprinkling of Chinese . . . day labor averaging $1 per day, while contract labor, of course, amounted to more, according to conditions.*
>
> *Fourteen months only were required to build these twenty-four miles of tunnels and open ways, with flumes spanning deep gorges, hundred of Japanese being employed on the work. They bored in from the upper sections, near the intakes, and bored in from the mouth. Along the trail at stated intervals crosscuts were driven into the tunnel line, and two gangs started at each crosscut, working out from each other, until at one time gangs were working on eighty different breasts. These crosscuts formed the windows, so to speak, of the underground system, openings through which the excavated material was drawn and cast out over the pali, and it was through these same openings that the materials were passed after being laboriously carried on the backs of mules who trod the trails from headquarters, carrying powder, fuse, lumber, cement, food, and all supplies. It was from these openings that the Japanese laborers emerged when their work was completed, extinguished their lanterns and candles, and climbed to their nest-like shelters, perched on perilous points where their wives and children lived during the day, and babies clambering up and down the dizzy trails and to and from the houses.[6]*

The reference to families in the camps was inconsistent with common practice and perhaps was untrue. Less surprising was "the comfortable home occupied by Jorgen Jorgensen and his mother, where guests are put up and refreshed."[7]

It was reported that such accommodations reflected McCrosson's philosophy that comfort, coffee, and good cheer on the job were good business.

There were few problems on the job, noted William Payne, in charge of the tunneling: "The tunnel men struck work for a few days, but by giving them a light raise in their contracts, they are now well satisfied." Payne took pride in his final report that all tunnels were built without an accident. The only death he reported during construction was that of an engineer named Thomas Kelly who, returning home one night, drowned when he was swept off his horse while crossing the river. He was buried up in the valley, where his headstone can still be seen at *haole make.*

Shortly before the opening ceremony, a huge boulder rolled down the slope and crushed the Kawainui intake. The local newspaper speculated: "To the mind of the Hawaiians who came to witness the ceremony of dedication, the thought came that the goddess of Waipio had made her final protest against the imprisoning of the waters of her domain by unloosening a boulder high up in her castellated home and with unerring aim hurled it directly into the center of the flume, crushing it like an eggshell." The report continued that Rebecca Kaukahi, on the other hand, offered a chant evoking the radiance of Kalakaua to thank the goodness of God who had brought the water of darkness to the light of the day and to thirsty Hamakua.[8]

The Lower Ditch was extended to service Paauilo in 1912 and then to Paauhau in 1918. It now ends at Reservoir 324B, built in 1974, at Paauilo. The tunnel from the dam to the reservoir in Lalakea gulch was built around 1920. Originally, very little water was developed from the Waima tributary of the Wailoa River, since its main source was the springs below the tunnel. These springs were tapped in 1966, under manager Richard Frazier, by means of two pumps that lifted the water up 380 feet to the tunnel. These springs were used only during drought conditions, however, as the cost of pumping was high. Although these pumps originally had a capacity of 6 mgd, by 1984 the flow had diminished to 5 mgd. A third pump was placed about 200 yards down stream to take advantage of Kawainui spring to supplement the stream water when necessary.

In 1915, Hamakua Irrigation Company leased the *ahupua'a* of Waipio (including the *kuleana* and the rice mill) from Bishop Museum, presumably to gain control of land that might be able to claim appurtenant or riparian rights to water that was being diverted out of Waipio by the two Hamakua ditches. Sixty-one acres were in rice on the valley floor. Hamakua Irrigation Company invested substantially in supporting the rice farmers by improving the irrigation system and the rice mill and providing marketing and distribution services.

Tunnels are enduring testaments to the men who built them. In addition, on the Waia-hole Ditch this rare monument announces "Project Completed, June 1916," followed by the names of contractor Mizuno, his surveyor, stonemason, and workers. (Photo: D. Franzen)

Rice had supplanted taro in windward valleys throughout Hawaii from the 1860s through the 1920s. "Sandwich Island rice" commanded top prices in San Francisco and for fifty years was the leading export after sugar, peaking in the late 1880s at 12 million pounds. The export and local market declined steadily after that time, however, and the Waipio farmers were left holding the bags when the rice market crashed in 1921. HIC reported for the year ending December 1921: "During the last six months of the year about 60 percent of the land formally planted to rice has been abandoned and is now lying idle."

Within a few years, only a small Hawaiian rice industry was left, mostly on Kauai, primarily for specialty mochi rice. The collapse of the Hawaiian rice industry was as much a product of its own failure to organize itself as it was a result of foreign competition. There was no unifying force—like the Big Five in the sugar industry—to analyze the market and respond to it, to ensure consistent quality, to improve technology, and to finance capital improvements or carry the planters through lean times. The success that sugar made look so easy was in fact achieved by an extraordinary degree of cooperation that was not duplicated in Hawaii's other agricultural enterprises.

While the building of the ditches went smoothly, the same cannot be said for their use. Disputes over water dominated the correspondence between plantations, HIC, and agencies, the agendas at the director's meetings, and the passions of those involved. At issue was water allocation, measurement and cost, ditch outlets, maintenance, surplus water, accounting practices. In 1909 the conflict moved to the courts and into the field. When Hawaiian Irrigation Company removed a measuring device critical to Honokaa Sugar Company, Honokaa Sugar employees shut the water off to Paauhau Sugar Plantation and tried surreptitiously to survey the reservoir on a Sunday. Caught in the act by Jorgensen and McCrosson, who ordered them to leave, they returned to finish the job when the two men were known to be elsewhere. As the Honokaa manager explained to agent F. A. Schaefer in September 1909: "As they were standing on the bank of the reservoir, one M. Sprout (employed by the Haw Irr. Co) came up with one axe and smashed Mr. Williamson's instrument all to pieces—then he walked away . . . and on Tuesday we sued Mr. Sprout for 'Malicious Injury.' " Later in the month Honokaa Sugar noted a new HIC ditch tender "who on several occasions had 'monkyd' with the outlet gates—hence we locked same."

Hawaiian Irrigation Company's troubles did not stop there. In 1922, the trustees of Bishop Museum, lessors of the "waters of Waipio," took strong exception to HIC's request for reduced lease rent based on the claim that it was losing money. The trustees questioned HIC's report and its bookkeeping procedures. A letter to HIC noted that the report *fails to disclose any reason why the Trustees should sell to you for $5,000 water which the Company sells for $150,000; should waive half of their rent to enable the Company to accumulate out of revenue produced by the sale of their water a sinking fund to retire the Company's bonds (the sale of which provided the funds to construct the ditches), and provide a fund with which to pay the stockholders $1,250,000 (of which $1,150,000 is represented by "franchises and contracts") at a future date.*

As previously noted, the sugar situation on the Hamakua and Kohala coasts was different from most other areas in one important respect. Whereas on most other islands large areas were controlled by one or two plantations—or several adjacent plantations cooperating under the unifying force of one agency—on the Hamakua and Kohala coasts any number of agents controlled numerous independent plantations, planters, and mills, each vying for predominance. Plantations included Halawa Sugar Company, Hamakua Mill Company, Hawi Mill & Plantation, Kohala Plant, Kukaiau Mill Company, Kukaiau Plantation Company, Laupahoehoe Sugar Company, Niulii Mill & Plantation, Paauhau Sugar Plantation Company, Pacific Sugar Mill, and Puakea Plantation Company. Agents included Schaefer & Company, Davies & Company, Hackfeld & Company, Irwin & Company, and Castle & Cooke. As a consequence of this diversity, the benefits of the cooperative effort found elsewhere in Hawaii were not enjoyed here. Both the Kohala and Hamakau ditch companies were independent water companies in the business of making money—as opposed to most other water companies, which were subsidiaries of their clients devoted to making money for them.

Although this system was highly competitive, some of the investors in the key plantations were also investors in the water companies—and with this combined leverage, these were the companies that succeeded in absorbing the other interests. The financial and personal relationships between the Hamakua Irrigation Company, the Honokaa Sugar Company, and Oahu's big money were close and complicated. Some of the various interests represented at the opening of the Lower Ditch were the "father of the project," John McCrosson, plus the following:

President Harry Lewis, the chief stockholder of the Hamakua Ditch Company, wife and son; R. W. Shingle, president of the Waterhouse Trust; Frank Thompson, attorney for the Ditch Company; W. G. Irwin, representing the bondholders of the Ditch Company; E. Faxon Bishop, representing the Brewer & Company plantations; L. A. Thurston, large stockholder and attorney for plantations in the neighborhood; Manager Gibb of Honolulu plantation [recently with Paauhau Plantation]; J.A.W. Waldron, of F. A. Schaefer & Company, agent for Honokaa and Pacific Sugar plantations; representatives of the Honolulu newspapers, a number of ladies and practically everybody from the surrounding plantations.[9]

The conflicts between Honokaa Sugar Company and HIC were resolved through a takeover of both companies by J. W. Waldron in January 1915. This was unlike the takeover of Hawaiian Sugar Company by Olokele Sugar on Kauai, where the new management was indeed entirely new. A review of the Hamakua coast company's records suggests that this was not so much a takeover as a reorganization with the purpose of consolidating the major companies under the agency of Schaefer & Company. The old Board of Directors of HIC comprised R. W. Shingle, J. T. McCrosson, E. I. Spalding, F. E. Thompson, A. A. Wilder, and E. H. Wodehouse. The new HIC Board consisted of J. W. Waldron, G. E. Schaefer, and W. Lanz, while McCrosson and Shingle became directors of Honokaa Sugar Company. The HIC agency agreement was transferred from Davies to Schaefer & Company, the agent and major stockholder of Honokaa and Pacific Mill. While at first Waldron served as manager of HIC without salary, a few years later he was president of HIC, secretary of the Board of Directors of Schaefer and Company, and president of both Honokaa Sugar Company and Pacific Mill. His wife was Schaefer's daughter.

Further consolidation took place in 1928, when Honokaa Sugar and Pacific Mill formally merged (although they had been under the same management since 1913). It must have been confusing for the management of the various companies: while on the one hand they had to be competitive to maximize profits, on the other they had to cooperate with each other because that was the model that had always succeeded. The dilemma is suggested by a nasty letter written by Waldron from himself (on behalf of Hawaiian Irrigation Company) and to himself (on behalf of the Honokaa and Pacific Mill plantations) demanding higher rates for the water.

By 1930, Schaefer & Company owned a key portion of the Hamakua coast. But Davies and Company, agent for Hamakua Mill Company and Laupahoehoe Plantation, had bigger plans. It started negotiations to acquire Honokaa Sugar in 1936. Finally, in 1952, Theo Davies bought Schaefer & Company and gained control of Honokaa Sugar and HIC.

Davies bought the Paauhau Sugar Company from C. Brewer in 1972. In 1979, the Davies Hamakua Sugar Company was formed from the merger of Laupahoehoe and Honokaa plantations. With over 35,000 acres of cane land, it was the largest sugar plantation in Hawaii. Unique logistics problems were presented because the new plantation was 35 miles long and just 4 miles wide. To minimize transportation costs between field and mill, two mills were maintained. In 1980, the Davies Hamakua Sugar Company bought the 3800-acre *mauka* watershed in Waipio from Bishop Museum, although the museum kept the Waipio *ahupua'a*. Hamakua's sugar plantations were the last to conform to

Flumes on the Kohala Ditch. As sugar companies closed, ditch maintenance was curtailed or ceased entirely. (Photo: D. Franzen.)

the mold of agribusiness, all other models proving to be uncompetitive and unprofitable.

On 1 January 1984, Francis Morgan stunned Hawaii's business community by buying out Theo Davies' sugar interest on the Hamakua coast. At a time when plantations verged on financial disaster, when one company after another was shutting down, when economists were predicting the demise of this industry in Hawaii, Morgan became the only sole owner of a sugar plantation in Hawaii. Morgan had anticipated peaks in sugar prices to finance his purchase. The price of sugar remained flat, however, and the company was unable to survive. In 1993, Hamakua Sugar Company, declared bankruptcy, leaving the banks, the unions, and the state as major creditors.

The state had more than the usual health, safety, and welfare concerns in the bankruptcy, as several years earlier it had loaned Hamakua Sugar $10 million in a futile attempt to avert the company's collapse and thus was a creditor as well. The County of Hawaii was involved in the discussions regarding future land planning. Even the federal government played a role: the Department of Defense and the Department of Agriculture's Natural Resource Conservation Service funded planning and transition efforts for this region—in particular, for the badly deteriorated Lower Hamakua Ditch.

There was competition for the newly available Hamakua water. Both the "upside" farmers and the Waipio valley farmers claimed a need for it, although in some cases the claim was not substantiated by realistic farming ventures. Many advocated stream restoration, which was supported by the Hawaii State Constitution "where practicable." A key issue was the need for cheap water for future urban development—but since the land was not zoned for this kind of development and such plans often provoked local resistance, these concerns remained largely unspoken. There were concerns, as well, that the ditch system could never be restored if it were to fall into total disrepair—for reasons of expense, loss of rights-of-ways, and politics—and therefore should be maintained to keep future options open. Of primary concern, of course, was who would pay. Many turned to government to shoulder the financial burden. Others urged government to set priorities first, to formulate criteria, and to develop a plan. In short, most of the issues surrounding Hamakua Sugar Company's bankruptcy and transition were the very same ones surrounding other Hawaii communities and their ditches in the wake of the failing sugar companies.

APPENDIXES

Appendix 1: Letter from the Attorney General (1876)

Letter from Attorney General William R. Castle to His Excellency Wm. L. Moehonua, Minister of the Interior, dated 7 September 1876:

Sir:

The application of Messrs Castle and Cooke, representing the Haiku Sugar Company, Alexander and Baldwin, James M. Alexander, the Grove Ranch Plantation and Capt. Thos H. Hobron, dated August 21,1876, has been placed before me. This application requests permission to take water from several streams, in Koolau Maui, to be carried to their respective sugar plantations, for purposes of irrigation.

So far as I am informed, this application is new, in its nature. It is not for land, nor, as I understand, for an absolute sale or grant of the waters of the streams mentioned in the application. The application is for a license; the license to take and use water, conveying the same in part over several government lands.

Several questions are suggested upon the matter, of which the more important appear to be:

1st. Has the Minister of Interior the power and authority to make such a grant?

2nd. What can be conveyed or granted? &

3rd. Is the use asked by the application contemplated by our law?

Upon these questions, my opinion is as follows.

The great act of 1848, confirming the gift of lands to the people, as made by Kamehameha III, confers upon the Minister of the Interior, full power to direct, superintend and dispose of said lands; provided however that the terms and conditions of sale should be approved by the King in Privy Council. No

163

further provision or restriction was made by law till the year 1859 when the Civil Code was published as then revised and became the law of the land. Sections 39 to 48 inclusive refer to government lands and kindred property. Section 42 expressly provides that "by and with the authority of the King in Cabinet Council" he shall have "power to lease, sell or otherwise dispose of the public lands and other property in such manner as he may deem best for the promotion of agriculture and the general welfare of the kingdom, subject however to such restrictions as may from time to time be expressly provided by law." By the terms of section 40, "streams, ponds, springs, watercourses" &c constitute part of such property. Section 48 prohibits said Minister from disposing of certain springs and ponds near Honolulu, and all other government water ponds, springs and streams "which may be valuable for public use." By the laws of 1874, chapter 24, the Minister is prohibited from selling any land the value of which is over $5,000 without the consent of the King in Privy Council. No other laws have been passed affecting the question. Subject therefore to the provisions of these sections read together, the Minister of the Interior has full power to make the grant asked, for it will be seen that the law of 1874 does not apply. For the purposes of the case in hand, it needs only the consent of the King in Cabinet Council. It may be claimed that the provisions of section 48 prohibit the disposal of the water asked for. The answer to this is—that as there are no cities, towns or villages, and at best but a very sparse population in that region and the waters from time immemorial run waste into the sea there can be no public use for which they are so valuable as to prevent a disposal. In addition to which, the provisions of section 48 probably apply only to absolute alienations of title.

The application asks for the right to take water. It asks a grant of a license which may be made by deed of lease.

The answer to the third question seems to me to be very clear. It is asked that water be taken for the purpose of irrigation, in short—for the uses of agriculture—an interest particularly specified as one which the government should foster and encourage and for which a disposal of the public property may be made. The Reciprocity Treaty having passed and a brighter future opening for the country, it becomes the duty of the Government to aid and foster in every possible way the agricultural interests of the country upon which our prosperity mainly depends. In offering and furnishing such aid anything like a monopoly must be guarded against. The government acts for all parties and should endeavor to distribute equally whatsoever of favor it may have, or, as in this case—when no favor is asked but a license is requested—to guard against any injury to private rights by the establishment of any monopoly. At some future day the government—as is the case in some of the European nations—may undertake the work of carrying water from place to place as the country may need, but at present is not prepared to engage in any such development of internal resources, and for such water may demand a reasonable compensation. Its ponds, springs and streams are valuable and should be guarded and protected. Until the government is ready to undertake such work—no obstacle should be thrown in the way of others, who are able and ready to commence such work. It seems a fit and proper thing to grant the application made by Messrs Castle and Cooke, reserving certain rights to the government, as will be hereinafter specifically set forth. The applicants propose to begin work immediately, the (*sic*) have already gone to considerable expense—as I am informed in surveys &c. At present, of course, the work is largely experimental—if successful it will largely increase the value of their lands—as well as those adjoining. In this the whole nation will join indirectly and it is but just that very liberal terms should be made. It is desired to begin at once. For the purpose of allowing that, I would suggest that an answer be sent to Castle and Cooke, immediately containing the following conditions and terms.

The Government will grant to the Haiku Sugar Company, Alexander and Baldwin, James M. Alexander, the Grove Ranch Plantation and Captain Thos H. Hobron and their respective successors, heirs and assigns, the license to take water from the streams named in the application and to carry the same over all government lands intervening between the said streams and the remotest land to which it is *now* desired to carry said water, for the period of twenty years from date of acceptance of these terms, at an annual rent of one hundred dollars, *Upon condition* 1st That a sufficient ditch, canal or other waterway shall be at once commenced and finished in a reasonable time. 2nd That this grant shall not interfere with the rights of tenants upon said lands or streams. 3rd nor shall it in any way affect the right of the government to grant to any person or persons the right to take water (not to interfere with the water hereby granted) from the same or other streams to be carried over the same land or lands for any purpose whatsoever, and if need be, to be carried through the ditch, canal or other waterway to be constructed by these grantees, provided however, that during the said period of twenty years the supply of water, a right to take which is hereby granted, shall not be diminished by act of the government, and 4th That at any time during said period the government may purchase the said canal, ditch or other waterway upon payment of the actual cost thereof only, and in case of such purchase, will continue to furnish water to these grantees at a just and reasonable rate not to exceed that paid by other parties taking water from such ditch or other waterway.

In case such a communication is sent and accepted, all rights of the government will be reserved, the work can begin at once, and the necessary deeds drawn up in accordance with the above terms at a more convenient day.

I return herewith the application of Messrs Castle & Cooke.

I am sir most respectfully yours.
Wm R. Castle
Attorney General

Appendix 2: Hydroelectricity

In the early days of sugar, due to the high cost of coal the cost of pumping groundwater was prohibitive. Hydroelectric development was one of the many new technologies of the industrial revolution, and Hawaii's plantations were eager to develop this cheap power. Though most of these powerplants were small, the cumulative power they produced was significant—and this was supplemented by power generated at the mills through burning bagasse. Many plantations were able to provide for all their own power needs, which included running the pumps and factory and powering the communities, and some even had surplus power to sell.

One of the earliest plants was built in Waianae. In 1897, Waianae Plantation's John Dowsett built a reservoir and then installed a hydroelectric plant capable of generating 300 kilowatts: "This electric power drove the plantation's water pumps at the wells, operated the mill generators in the off season and provided electric lights for the plantation manager's house at a time when many people in Honolulu were still using kerosene lamps."[1]

Hydroelectricity is typically generated by instream dams and power generators. In this type of system, water is usually stored behind a dam and energy production is managed by regulating the flow through the dam. Hawaii's conditions call for a different design, as its geography does not lend itself easily to storage. In Hawaii's "run-of-the-river" plants, energy is produced by utilizing water's rapid drop in elevation from the mountains to the sea. Water is diverted from the stream at higher elevations, directed with as little loss of elevation as possible to a forebay (a confined pool not affected by surges), then dropped through a penstock (a pressurizing pipe) to the off-river powerplant.

Most of the early hydroelectric plants in Hawaii utilized the Pelton waterwheel, designed by American engineer L. A. Pelton. The amount of energy produced is determined by the quantity of water, the drop (the vertical distance between forebay and powerplant), and the efficiency of the powerplant. These run-of-the-river designs are not considered to be firm power sources, as their energy production rises and falls with the flow of the river. After the water leaves the powerhouse, it is in some cases returned to the stream; in others it is sent directly to irrigation.

While a cursory inventory reveals more than twenty hydroelectric plants built in conjunction with the plantation irrigation systems, there were in addition numerous small plants at the camps and the mills. There may have been a hundred and more of these, scattered throughout Hawaii, used for small jobs such as running lights or small machines. At Pioneer Mill, for example, "dozens" of smaller hydros were reported at the various camps; on the Hamakua coast, almost every mill had a small hydro plant. Sometimes a hydroelectric plant was put on line exclusively to support the ditch's construction, as at Waiahole and in Lahaina.

The 1906 McBryde's Wainiha Powerplant was the first hydroelectric installation in Hawaii with significant power production—to this day it produces more output annually than any other single plant. When all of its powerplants are taken together, HC&S has the largest total power production capability. Of the eighteen hydroelectric powerplants operating in Hawaii today, eleven were built by plantations. In 1990, hydroelectricity supplied about 1.5 percent of the state's total electrical energy consumption.

On Kauai, Lihue Plantation built the Waiahi Powerplant on the south fork of the Wailua River in 1914. The system took two years to complete under the direction of Hans Isenberg. It had two Westinghouse generators with Pelton waterwheels capable of generating 600 kilowatts. The power was used for mill operations, pumping irrigation water, and employee needs. The Waiahi Electric Company was set up as a public utility servicing four small towns. In 1931, the powerplant capacity was increased to 800 kilowatts using a Francis turbine. In that same year a 500-kilowatt plant was placed further upstream at the 1050 foot elevation. This upper Waiahi Powerhouse uses a Pelton waterwheel and a G.E. generator.

McBryde Sugar Company's 1906 Wainiha Powerplant and 1928 Kalaheo Powerplant are discussed elsewhere in the book.

Koloa Plantation installed a 120-kilowatt hydroelectric powerplant on Koloa stream in 1918, taking advantage of the water from the Waiahi–Kuia aqueduct. It was abandoned in 1965.

Hawaiian Sugar Company installed a 500-kilowatt hydroelectric plant in 1921 on the Olokele Ditch to take advantage of a drop of 228 feet. Although still in place, this hydroplant has not operated much since Olokele Sugar Company put in a more efficient 1250-kilowatt plant parallel to it in 1982.

Pelton waterwheel buckets, Waiawa Power Station, Kekaha Sugar Co. Kauai, 1908. The Pelton waterwheel was a major technological development in hydroelectric production. Although refined later, the basic design remains one of the most efficient ways of producing energy from water. (Photo: D. Franzen.)

Kekaha Sugar Company installed the 550-kilowatt Waiawa Powerplant in 1908 at a cost of $64,500, utilizing a 280-foot drop on the Kekaha Ditch. Four to five hundred more acres above the ditch were put into cane, irrigated with the help of this new plant to pump water to the higher elevation. This early powerplant retains much of its original equipment. The 1908 Pelton waterwheel remains, although the buckets and nozzles have been replaced. The transformers have mechanical switches, and some of the lettering on the old switchboard is by hand. The plant is one of only a few in the islands that was still being placed on line by hand in the 1980s.

Kekaha Sugar's Mauka Powerhouse was completed in 1913, at a cost of $170,000, with a 1200-kilowatt capability. The collection system consisted of twenty-three tunnels, the longest being 1060 feet, and a total length of 13,000 feet. This plant was modernized in 1930, destroyed in a flood in 1949, rebuilt in 1952–1954, and overhauled in 1981. Its present capacity is 1000 kilowatts. It was damaged by Hurricane Iniki on 11 September 1992 and restored to operation the next year.

On Oahu, Waialua Sugar Company built a 450-kilowatt hydroelectric plant at Kaukonahua in 1916. Reportedly it was almost impossible to have a lone operator stay up there at night on account of the ghosts: it was two men or nothing. As it would have cost too much to automate for the small amount of power the plant produced, it was closed in 1960.

On Maui, HC&S had three hydroelectric plants, all utilizing water collected by the EMI irrigation system. The earliest, Paia Hydro, was built by Maui Agricultural Company in 1912 with a 800-kilowatt capacity. In 1923, the penstock was extended to a higher elevation, thus increasing the capacity to 1000 kilowatts. HC&S built a 4000-kilowatt hydroplant at Kaheka in 1924. In 1982, a 500-kilowatt hydroelectric powerplant was installed at the Hamakua Ditch above Paia. Located only 50 feet below the Wailoa forebay, this "low-head" hydroplant takes water through a 36-inch pipe and discharges it into the Hamakua Ditch.

Besides these three hydros, HC&S has a bagasse-powered steam powerplant at the Paia factory, and the Central Powerplant, built in 1918, located at Kahului. In 1921, electric lighting was brought to the camp houses. By the 1930s this was the largest plantation power system in Hawaii, with a 12,000-kilowatt capacity. The largest consumer was the water pumps (6000 kilowatts), then the factory (1500 kilowatts), and general uses such as lighting, feed mill, dairy, carpentry shop, refrigerator plants, machine shops, and "talkie movie houses" (400 kilowatts). Surplus power (900 kilowatts) was sold to Kahului Railroad Company and to Maui Electric Company. The Central Powerplant supplied power for all of central Maui until after World War II. In 1984, the combined total capacity of all HC&S power-generating systems was rated at 37,300 kilowatts.

On West Maui, a hydroelectric plant was installed in 1912 at Honokohua gulch by Honolua Ranch (later Maui Land & Pineapple Co.) specifically to provide power to camp and equipment during construction of the Honolua and Honokowai tunnels. By 1931, Pioneer Mill had four hydroelectric plants and one oil-fired steam powerplant at the factory. Power was distributed for irrigation pumping (4200 kilowatts), for the factory (600 kilowatts), and for miscellaneous uses such as shops, rock crushers, wharf machinery, dairy plant, feed cutter, and camp lighting (500 kilowatts). The remainder (350 kilowatts) was sold to Baldwin Packers pineapple cannery and the Lahaina Ice Company, which until the early 1960s was the local public utility. By 1984, only one hydro remained: a 300-kilowatt powerhouse dating from 1912.

On Hawaii, there were at least four small hydroelectric plants in Kohala built by the various plantations. One of these, in Hawi, was modernized in 1935, increasing capacity from 175 to 350 kilowatts; it used 10 to 12 mgd. There was another hydro at Union Mill. These have all been abandoned, the last in 1976. However, the Hawi Agricultural & Energy Corporation took advantage of one of the original penstocks to build a new powerplant with two inductive generators of 170 kilowatts each. And, on the other side of the Kohala Mountains, a 800-kilowatt hydro was built at Haina in 1945. In 1995, the Haina mill was sold along with the hydro and moved to Japan.

Glossary of Hawaiian Words

Definitions are from Hawaiian Dictionary, *revised and enlarged edition by Mary Kawena Pukui and Samuel H. Elbert (Honolulu: University of Hawai'i Press, 1986).*

ahupua'a: Land division usually extending from the uplands to the sea.

ali'i: Chief, chiefess, king, queen, noble; royal, kingly, to rule or act as a chief, govern, reign; to become a chief.

'auwai: Ditch.

hā: To breathe, exhale; to breathe upon; breath, life. Also: stalk that supports the leaf and enfolds the stem of certain plants such as taro, sugarcane; layers in a banana stump.

'ili: Land section, next in importance to *ahupua'a* and usually a subdivision of an *ahupua'a*.

kalo: Taro *(Colocascia esculenta)*, cultivated since ancient times for food, spreading widely from the tropics of the Old World. In Hawaii, taro has been the staple from earliest times to the present, and here its culture developed greatly, including more than 300 forms.

kānāwai: Law, code, rule, statute; legal—perhaps so called because many early laws pertain to water *(wai)* rights.

kō: The sugarcane *(Saccharum officinarum)*, a large unbranched grass brought to Hawaii by early Polynesians as a source of sugar and fiber.

konohiki: Headman of an *ahupua'a* land division under the chief; land or fishing rights under control of the *konohiki*.

kuleana: Right, title, property, portion; a small piece of property, as within an *ahupua'a*.

lo'i: Irrigated terrace, especially for taro.

makai: At the sea, seaward.

maka'āinana: Commoner, populace, people in general, citizen.

mauka: Inland, at the mountains.

pani wai: Dam, sluice, levee, dike.

wai: Water, liquid of any kind other than seawater, juice, sap, honey; any liquid discharged from the body, as blood, semen; color, dye, pattern; to flow like water; fluid. *Wai* has a connotation of wealth and life.

waiwai: Goods, property; value, worth; estate; rich, costly; financial.

Notes

Introduction

1. Hawaii Stream Assessment (1991). Out of a total of 270 perennial streams, 28 have a main stem of over 10 miles. Eleven have median flows greater than 30 mgd; another forty or so have median flows greater than 6.5 mgd.

1. Pioneers, Politics, and Profits

1. "Indenture Between Hawaiian Kingdom and Claus Spreckels," July 1878.

2. Jorgensen Report on Upper and Lower Hamakua Ditches, with letter from P. H. Bartels, ca. 1921. A rare instance of federal government interest in developing surface water in Hawaii occurred in 1939 when the U.S. Bureau of Reclamation proposed developing an aqueduct on windward Molokai from Wailau Valley through Pelekunu and Waikolu valleys to service some 16,000 acres of Hawaiian Homes agricultural lands on central Molokai. It was proposed to use national defense money because "some 500 or more families can be supplied with farms and . . . the result will be to further the rehabilitation of the Hawaiian race, and make the Islands less dependent on Mainland importation of foodstuffs, thus contributing to national defense." Little more was ever heard of this project.

3. Hutchins, *Hawaiian System of Water Rights*, p. 18.

4. Gilmore, *Hawaii Sugar Manual* (1932), p. 5.

2. Water Use and Rights

1. *Reppun v. Board of Water Supply* (1982), p. 546.

2. Ibid., p. 540.

3. *Beckley v. Ohule* (1873).

4. *Cartwright v. Gulick* (1886).

5. Letter from Spreckels to Kalakaua, 5 Dec. 1878.

6. The only known exception was a small but historically significant diversion at Waiapuka, Kohala, attributed to King Kamehameha. Here a tunnel passes through a small ridge to bring water over from a wetter valley to irrigate taro *lo'i*. A series of vertical shafts extend from the land surface to the tunnel roof, perhaps to guide the tunnel's path during construction. See *Thrum's Hawaiian Annual 1919*, pp. 121–126, "A Little Known Engineering Work in Hawaii" by J. N. S. Williams. Besides being attributed to Kamehameha, tradition also has it as "menehune tunnel," and it is also thought by some to have been built by a Mr. [Samuel?] Parker.

7. *Reppun v. Board of Water Supply* (1982), p. 546.

8. *Horner v. Kumuliilii*, 10 Haw. 174. 176 (1895).

9. Letter from Daniels to Hutchinson dated 23 April 1866.

10. Bushnell, O. A. *The Gifts of Civilization*, p. 268.

11. *Reppun v. Board of Water Supply* (1982), p. 544.

12. A search of the Hawaiian language record of the 1850–1900 period would contribute to a greater understanding of this subject.

13. Cooper, "A Political and Legal History of Water Rights in Hawaii's Streams," p. 64.

14. *Territory v. Gay*, 31 Haw. 376, 393 (1930).

15. There are various and often conflicting interpretations of the events. Interested readers are advised to refer directly to the court decision.

16. *McBryde v. Robinson* (1973), pp. 175–187.

17. Van Dyke et al., *Water Rights in Hawaii*, p. 145.

18. *McBryde v. Robinson* (1973), p. 175.

19. *Reppun v. Board of Water Supply* (1982), p. 545.

20. Ibid., p. 555.

21. Tuttle Report (1902).

22. *Environmental Defense Fund Letter* 25(3) (1994): 1.

23. American Rivers, 1993 Annual Report, Chairman's Message, p. 3.

24. Article XI, Section 1, Constitution of the State of Hawaii.

25. Article XI, Section 7, Constitution of the State of Hawaii.

26. Hawaii Revised Statute, chap. 174C.

3. The Ditch Builders

1. Edward D. Beechert's *Working in Hawaii: A Labor History* (1985) provides a full account of Hawaii's labor history.

2. O'Shaughnessy, "Irrigation Works in the Hawaiian Islands," p. 460.

3. Herschler, *50 Years of Water Service*, p. 12.

4. Early Efforts

1. Damon, *Koamalu*.

2. *Wailuku Sugar Company Centennial*, p. 2.

3. Letter from Alexander to Wilcox dated 25 Sept. 1876.

4. FLC, "The Hamakua-Haiku Irrigation Ditch," pp. 39–40.

5. Dean, *Alexander & Baldwin, Ltd. and the Predecessor Partnerships*, p. 19.

6. FLC, "The Hamakua-Haiku Irrigation Ditch," p. 40.

7. Adler, *Claus Spreckels: The Sugar King in Hawaii*, p. 49.

8. Ibid., p. 49.

9. Letter from Alexander to Wilcox dated 25 Sept. 1876.

5. East Kauai

Koloa Plantation

1. Although O'Shaughnessy's obituary in the *Honolulu Advertiser* reported that he engineered the Koloa Reservoir, I found no other reference to substantiate that claim.

McBryde Sugar Company

1. McBryde Sugar Co. field notes, 20 Sept. 1927 and 12 May 1928 (McBryde files).
2. Stearns, *Ground Water Supplies for Pioneer Mill Co.*, p. 51.

6. West Kauai

Hawaiian Sugar Company

1. O'Shaughnessy, "Irrigation Works in the Hawaiian Islands," p. 160.

Kekaha Sugar Company

1. Letter from A&B to Hackfeld & Co. dated 1906 (Kikiaola Land Co. files).

7. Oahu

Waiahole Water Company and Oahu Sugar Company

1. An example of how statistics can vary. These are from an Oahu Sugar Co. hand-out; Herschler reported 25.3 miles; Thrum reported 26.4 miles.

2. Bishop, Progress Report on Koolau Tunnels, Feb. 1913.

3. Jorgensen, "Monthly Reports to Waiahole Water Co.," Dec. 1913.

4. Kluegel, "Engineering Features of the Water Project of the Waiahole Water Co.," pp. 93–107.

5. Ibid., p. 105.

6. Ibid., p. 176.

7. *Thrum's Hawaiian Annual 1895*, p. 144.

Waialua Sugar Company

1. Smith. "Waialua Ag. Co. Has Elaborate Irrigation System."

8. East Maui

East Maui Irrigation Company

    1. Thayer, "The Lowrie Canal," pp. 154–161. A&B has provided figures that differ as follows: total tunnels 20,940 feet, total flumes 2210 feet, total siphons 4650 feet.

    2. O'Shaughnessy, "Irrigation Works in the Hawaiian Islands," p. 161.

    3. Ibid., p. 160.

9. West Maui

Wailuku Sugar Company

    1. *Wailuku Sugar Company Centennial: 1862–1962*, p. 28.

Honolua Ranch and Pioneer Mill Company

    1. O'Shaughnessy later reported this ditch to be 13.5 miles long.

    2. ML&P files, Manager's Report dated 31 March 1913.

    3. Pioneer Mill files.

    4. Ibid.

    5. Ibid.

    6. Ibid.

    7. Ibid., Kauaula Tunnel.

    8. Ibid., Kauaula Tunnel.

    9. Ibid., Kauaula Tunnel.

    10. All figures for average daily flow, unless otherwise indicated, are taken from Hatton, *Hydromania*, and are calculated for the twenty-year period 1956–1975.

10. Hawaii

Kohala Ditch Company

    1. Tuttle Report (1902).

    2. Kohala Sugar Co. Diary (1904).

    3. Letter from Robert Hall, Kohala Sugar Co. files (Bishop Museum).

    4. O'Shaughnessy, "Irrigation Works in the Hawaiian Islands."

    5. *Thrum's Hawaiian Annual 1907.*

    6. O'Shaughnessy, "Irrigation Works in the Hawaiian Islands," p. 464.

    7. Ibid., p. 461.

    8. *Thrum's Hawaiian Annual 1907.* In a personal communication, Bill Sproat recalls that seventeen men were lost during this project.

    9. Ibid., p. 117.

    10. Randolf, *Report on State Owned Waters and Water Facilities at North Kohala, Hawaii.*

Hawaiian Irrigation Company

1. There is a confusing earlier reference on a 1907 contract to "Hawaiian Irrigation Company, Ltd.," with McCrosson as vice president and E. I. Spalding as treasurer.

2. Jorgensen, "Report on Upper and Lower Hamakua Ditches."

3. HIC, Annual Report, 1915.

4. Jorgensen, "Report on Upper and Lower Hamakua Ditches."

5. *Hilo Tribune*, "Hamakua Ditch Is Now at Work," 5 July 1919.

6. Taylor, "Hamakua Ditch Marks New Era in Hamakua," *Pacific Commercial Advertiser*, 3 July 1910.

7. Ibid.

8. Ibid.

9. *Hawaii Herald*, "Opening of Great Hamakua System," 7 July 1910.

Appendix 2: Hydroelectricity

1. McGrath et al., *Historic Waianae*, p. 75.

Bibliography

I researched this project for the most part in 1984 and 1985. Most of my primary sources were found in plantation and sugar factor records. I have also relied on *Thrum's Hawaiian Annual*, Gilmore's *Hawaii Sugar Manual*, and Hawaiian Sugar Planters' Association publications for sugar and ditch statistics. DBED's annual publication, *The State of Hawaii Data Book: A Statistical Abstract*, provides additional information about sugar acreage, production, and freshwater use by island and type.

The United States Geological Survey (USGS) records and maps have been an invaluable reference. The drainage maps reproduced here are taken from the USGS's *Water Resources of Hawaii, 1909–1911* (1913). They show the status of surface water development by 1911 as well as the proposed Waiahole Ditch. Hawaii's ditches are on USGS topographic maps, which are updated periodically. The USGS publishes its water monitoring data annually in *Water Resources Data: Hawaii and Other Pacific Areas*.

References

Adler, Jacob. *Claus Spreckels: The Sugar King in Hawaii*. Honolulu: University of Hawai'i Press, 1966.

Alexander, Arthur C. *Koloa Plantation: 1835–1935*. Honolulu: Honolulu Star-Bulletin, 1937.

Alexander, W. P. *The Irrigation of Sugar Cane in Hawaii*. Report for HSPA. 1920.

American Rivers. Annual Report. 1993.

"Ancient Hawaiian Water Rights." *Thrum's Hawaiian Annual 1894*, pp. 79–84.

"A New Sugar Enterprise: The Hawaiian Sugar Co." *Planters Monthly* 8 (Nov. 1889): 485.

Baldwin, Arthur D. *Henry Perrine Baldwin: 1842–1917*. Private printing. Cleveland, 1915.

Beechert, Edward D. *Working in Hawaii: A Labor History*. Honolulu: University of Hawai'i Press, 1985.

Bushnell, O. A. *The Gifts of Civilization: Germs and Genocide in Hawai'i*. Honolulu: University of Hawai'i Press, 1993.

Collins, A. W. "Maui Agricultural Co." *Hawaiian Planters' Record* 8 (Nov. 1913): 11.

Cox, Doak. "Water Development for Hawaiian Cane Irrigation." *Hawaiian Planters' Record* 54(4) (1954): 175–197.

Cox, Joel B. "Water and Hawaiian Agriculture." *Paradise of the Pacific* (April 1942).

CWRM, DLNR, State of Hawaii. *Hawaii Stream Assessment: A Preliminary Appraisal of Hawaii's Stream Resources*. Honolulu: Commission of Water Resource Management and National Park Service, CPSU, 1990.

Damon, Ethel. *Koamalu*. Privately printed. Honolulu, 1931.

DBED. *The State of Hawaii Data Book: A Statistical Abstract*. Honolulu, 1994.

Dean, Arthur L. *Alexander & Baldwin, Ltd. and the Predecessor Partnerships*. Honolulu: Alexander & Baldwin, 1950.

"Ditches, Tunnels and Maui Wells." *Ampersand* (Spring 1979).

DLNR. *Value of State Owned Land of East Kauai*. Honolulu, 1963.

———. "Report on Estimated Costs of Operating the Waimanalu Irrigation System." Honolulu, 1978.

DOWALD. *Preliminary Report on the Waiahole Water Supply Project to Karl Singer*. Honolulu, 1947.

———. *The Fair Rental Value of State Owned Rights of Way, Honokohau Ditch, Honokowai Ditch, near Lahaina, Island of Maui*. Report prepared by Herschler and Randolf, consulting engineers. Honolulu, 1961.

———. *A Brief Description of the Upper Hamakua Ditch, Waimea, Island of Hawaii*. Circular C18. Honolulu, 1963.

———. *Waimanalo Irrigation System*, Report prepared for DLNR. Honolulu, 1977.

———. *Geology and Ground-water Resources of the Island of Kauai, Hawaii*. Bulletin prepared for DOWALD by G. A. MacDonald, D. A. Davis, and D. C. Cox. Honolulu: DLNR/DOWALD and USGS, n.d.

"EMI Is 100." *Ampersand* (Summer 1976).

Environmental Defense Fund Letter 25(3) (May 1994).

"Ewa Plantation Co." *Paradise of the Pacific* (Dec. 1919).

"EA." *San Francisco Chronicle*, 18 July 1910.

FLC. "The Hamakua-Haiku Irrigation Ditch (Maui)." *Thrum's Hawaiian Annual 1878*, pp. 39–42.

Gartley, A. "The Wainiha Electric Power Plant." *Thrum's Hawaiian Annual 1908*, pp. 141–158.

Gilmore, A. B. *Hawaii Sugar Manual 1931–1932*. New Orleans: HSPA, 1932.

———. *Hawaii Sugar Manual 1934–1935*. New Orleans: HSPA, 1936.

———. *Hawaii Sugar Manual 1938–1939*. New Orleans: HSPA, 1939.

"Hamakua Ditch Is Now at Work." *Hilo Tribune*, 5 July 1910.

"Hamakua Ditch Opening." *Thrum's Hawaiian Annual 1911*, pp. 138–142.

Hatton, Bert. "Hydromania: A Pioneer Mill Water Source Primer." Report for Pioneer Mill Co. Honolulu, 1976.

Hawaii Herald. "Opening of Great Hamakua System." 7 July 1910.

Hawaiian Irrigation Co. Annual Reports. 1915–1925. (Bishop Museum)

Hawaiian Sugar Planters' Association. *Hawaii's Sugar Islands.* Aiea, Hawaii, 1979.

Hawaii Tribune. "Hamakua Ditch Tunnel Completed." 21 June 1910.

Herschler, L. H. *50 Years of Water Service: The Waiahole Water Co.,* 1966.

Honokaa Sugar Co. Annual Reports.

Hutchins, Wells A. *The Hawaiian System of Water Rights.* Honolulu: Board of Water Supply, 1946.

Ikeda, Larry. "HC&S Centennial." *Ampersand* (Spring 1982).

"Jorgensen Obituary." *Honolulu Star-Bulletin,* 6 April 1944.

Kanalima Makahiki La Hoomanao O: 1898–1948. Kekaha Sugar Company's 50th year celebration publication. Honolulu, 1948.

Kauai Historical Society. *The Kauai Papers.* Lihue, Kauai: KHS, 1991.

Kekaha Sugar Co. Annual Reports. 1906–1930. (Kekaha Sugar Co. Files)

Kluegel, C. H. "Engineering Features of the Water Project of the Waiahole Water Co." *Thrum's Hawaiian Annual 1917,* pp. 93–107.

Krauss, Bob, and William P. Alexander. *Grove Farm Plantation.* Palo Alto: Pacific Books, 1976.

Lateef, Abdul. *A Study of Irrigation Water Management at Honokaa Sugar Plantation.* Oahu: HSPA Library, 1966.

"Lowrie Ditch." *Paradise of the Pacific* (August 1900): 10.

McCandless, James Sutten. *Development of Artesian Well Water in the Hawaiian Islands, 1880–1936.* Honolulu: Advertiser Publishing Co., 1936.

McGrath, Edward J., Jr., Kenneth M. Brewer, and Bob Krauss. *Historic Waianae: A Place of Kings.* Honolulu: Island Heritage, 1973.

"More Water Development, the Kohala Ditch, Wahiawa Dam and Reservoir, Other Projects." *Thrum's Hawaiian Annual 1907,* pp. 115–119.

"New Irrigation Works." *Thrum's Hawaiian Annual 1908,* pp. 156–158.

O'Shaughnessy, M. M. "Irrigation Works in Hawaii." *Thrum's Hawaiian Annual 1905,* pp. 155–164.

———. "Irrigation Works in the Hawaiian Islands." *Journal of Electricity, Power and Gas* 17(22) (1 Dec. 1906): 458–465.

Perry, Antonio. "Hawaiian Water Rights, Act of April 27, 1846." *Thrum's Hawaiian Annual 1913,* pp. 90–99.

Perry, H. C. "Ditch of the Hawaiian Sugar Company at Makaweli, Kauai." *Thrum's Hawaiian Annual 1892,* pp. 72–75.

Randolf, M. L. *Report on State Owned Waters and Water Facilities at North Kohala, Hawaii.* Report for DLNR. Honolulu, 1966.

"Retrospect for 1905: Water Development." *Thrum's Hawaiian Annual 1906,* p. 187.

Schuyler, J. D., and G. F. Allardt. "Irrigation." *Planters Monthly* 8 (Dec. 1889): 533.

"Securing the Wainiha Water Right Lease." *Thrum's Hawaiian Annual 1924,* pp. 95–112.

Smith, Jared G. "Science Etc. Ewa Plantation." *Honolulu Advertiser,* 24 Oct. 1923.

———. "Waialua Ag. Co. Has Elaborate Irrigation System." *Honolulu Advertiser,* 21 Nov. 1923.

Stearns, H. I. *Ground Water Supplies for Pioneer Mill Co.* Report for Amfac. Honolulu, 1964. (Amfac Office, Honolulu)

Sturgeon, George B. "Irrigation in Hawaii." *California Journal of Technology* 12(4) (Nov. 1908).

Sullivan, Josephine. *A History of C. Brewer & Company, Limited.* Boston, 1926.

Taylor, A. P. "Hamakua Ditch Marks New Era in Hamakua: Promoters Plan Much for Utilization of the Water in Addition to the Irrigation of Plantations." *Pacific Commercial Advertiser,* 3 July 1910.

Thayer, Wade W. "The Lowrie Irrigation Canal." *Thrum's Hawaiian Annual 1901*, pp. 154–161.

Thomas, W. B. "The Wahiawa (Oahu Ditch)." *Thrum's Hawaiian Annual 1903*, p. 73.

Tuttle, M. "To Trustees of Bishop Museum and Trustee Under the Will of Bernice Pauahi Bishop." Also referred to as the Tuttle Report (1902). (Hamakua Sugar Co., Pauilo, Hawaii)

USGS. *Water Resources of Hawaii, 1909–1911.* USGS, Territory of Hawaii. Water Supply Paper 318 by W. F. Martin and C. H. Pierce. Washington, D.C.: Government Printing Office, 1913.

———. *Geology and Ground Water Resources of the Island of Maui.* USGS, Territory of Hawaii. Circular by H. I. Stearns and G. A. MacDonald. Washington, D.C.: Government Printing Office, 1942.

———. *Water in the Kahuku Area, Oahu, Hawaii.* Water Supply Paper 1874 by K. J. Takasaki and Santos Valenciano. Washington, D.C.: Government Printing Office, 1969.

Vandercook, John W. *King Cane: The Story of Sugar in Hawaii.* New York: Harper & Brothers, 1939.

Van Dyke, Jon, et al. *Water Rights in Hawaii: Excerpts from Land and Water Resource Management in Hawaii.* Honolulu: State of Hawaii Dept. of Budget and Finance, 1979.

Wadsworth, H. A. "A Historical Summary of Irrigation in Hawaii." *Hawaiian Planters' Periodical* 37(3) (1933). (Reprinted in Gilmore, *Hawaii Sugar Manual,* 1936.)

———. *Index to Irrigation Investigations in Hawaii.* Honolulu: University of Hawai'i Press, 1949.

"The Waiahole Tunnel Project." *Thrum's Hawaiian Annual 1916*, pp. 174–180.

Waialua Agricultural Co. Annual Reports. 1902 to 1948.

———. Annual Report. 1948. (Summary on occasion of the fiftieth anniversary of Waialua Ag. Co., 1898–1948.)

Wailuku Sugar Company Centennial: A Century of Progress in Sugar Cane Cultivation: 1862–1962. 1962.

Waimea Sugar Mill Co. Annual Reports (Kikiaola Land Co., Waimea, Kauai)

Waldeyer, Carl. "Water Development: Mountain Tunneling." *Thrum's Hawaiian Annual 1910*, pp. 118–122.

"The Water Saga of Central Maui." *Ampersand* (Fall 1974).

Williams, J.N.S. "A Little Known Engineering Work in Hawaii." *Thrum's Hawaiian Annual 1919*, pp. 121–126.

Yardley, Paul T. *Millstones and Milestones: The Career of B. F. Dillingham, 1844–1918*. Honolulu: University Press of Hawai'i, 1981.

Unpublished

Alexander, S. T. Letter to G. N. Wilcox dated 25 Sept. 1876. (Grove Farm Homestead Collection)

Austin, Report covering state water supply on Kauai, 1962. (Amfac, Honolulu)

Bishop, H. K. Progress Reports on Koolau Tunnels for Waiahole Water Co., Feb. 1913 to Oct. 1913. (Oahu Sugar Co.)

Castle, William R., Attorney General. Letter to His Excellency W. L. Moehonua, Minister of the Interior, dated 7 Sept. 1876. (State Archives at Honolulu)

"The Commission of Private Ways and Water Rights." Vols. I and II. Puna (Manoa) district. (State Archives at Honolulu)

Cooper, George. "A Political and Legal History of Water Rights in Hawaii's Streams." Draft 1, 1978.

Daniel, Water Commissioner. Letter to Minister of the Interior F. W. Hutchinson dated 23 April 1866.

Davis, Rose Titus. "Major Streams of Kauai and Their Utilization." Master's thesis, University of Hawaii, 1960.

Furer, Fred. Letter to F. Weber, manager of Lihue Plantation, dated 1916. (HSPA LP files)

Grove Farm. Journals and ledgers. (Grove Farm Homestead Museum)

Hamakua Ditch Co., Ltd. Minutes from meeting of the board of directors, 1906–1909. (Bishop Museum)

Hawaiian Irrigation Co., Ltd. Directors' minutes, 1909–1915. (Bishop Museum)

Honokaa Sugar Co. Directors' minutes, 1913–1940. (Bishop Museum)

Jorgensen, Jorgen. Monthly reports to Waiahole Water Co., Nov. 1913 to Jan. 1916. (Oahu Sugar Co.)

———. Report investigating the Upper and Lower Hamakua Ditches for the Hamakua Sugar Co. with cover letter from P. H. Bartels, Assistant Supervisor, HIC, to H. W. Waldron, President of F. A. Schaefer & Co., 1921. (Hamakua Sugar Co. Pauilo, Hawaii)

Kangeter, John H., President, Engineering Association of Hawaii. Letter addressed to Kangeter dated 21 Nov. 1939.

Kehena Water Co. Account Book (Expenditures Only), 1912–1931. (Lyman House Museum, Hilo, Hawaii)

Kikiaola Land Co. Waimea Sugar Mill Co. files including: Correspondence regarding

Kekaha Ditch between Faye and Baldwin; Summary of Kekaha history written around 1920 when lease with territory was being renegotiated; Letter from George Mundon dated 1906.

Kohala Ditch Co. Ledger: 1905–1912. (Lyman House Museum, Hilo, Hawaii)

———. Trial Balance Book: Jan. 1920 to Dec. 1934. (Lyman House Museum, Hilo, Hawaii)

"Kohala Ditch Study." Draft 3, 10 Nov. 1959. (Amfac, Honolulu)

Kohala Sugar Co. Assorted documents. (Bishop Museum)

———. Diary of unidentified Kohala Plantation manager (likely Hart), 1899–1919. (Lyman House Museum, Hilo, Hawaii)

Makee Sugar Co. Letter to governor dated 14 Nov. 1911. (State Archives at Honolulu)

Maui Land and Pine Files. Honolua Ditch 1912–1913. Report with photos. (Maui Land and Pineapple Co. Files)

———. Plantation Manager's (Fleming's) monthly reports, 1912–1924. (Maui Land and Pineapple Co. Files)

McBryde Sugar Co. Managers' reports to the agents. (McBryde Sugar Co. Files)

———. *The Story of McBryde Sugar Co.: 1899 to 1949.*

Penhallow, David. Oral History of Bill Moragne. Kauai Museum, Lihue, Kauai, 1980.

Pioneer Mill Co. Files on Honokowai, Honokowai Bypass, Kauaula Tunnel Construction Notes, Honokohau Lining Notes. (Pioneer Mill Co. at Lahaina, Maui)

Schuyler, J. D., and G. F. Allardt. Report on Water Supply for Irrigation, Honouliuli and Kahuku Ranches. Oakland, Calif.: Jordan & Arnold, 1889.

Spreckels, Claus, and H. Schussler. Letter to King Kalakaua dated 5 Dec. 1878 regarding request for subterranean water. (State Archives at Honolulu)

Waimea Sugar Mill Co. Cash Book: 1900–1909. (Kikiaola Land Co., Waimea, Kauai)

———. Shareholders meetings and minutes, 1902. (Kikiaola Land Co., Waimea, Kauai)

Court Decisions and Hawaii Law

Beckley v. Ohule. Record of Private Ways and Water Rights 1873–1887, district of Kona, Oahu, p.14, decided 19 Nov. 1873.

Cartwright v. Gulick. Record of Private Ways and Water Rights 1873–1887, district of Kona, Oahu, p. 317, decided 18 March 1886.

Hawaiian Compiled Laws 1884. Honolulu: Hawaiian Gazette Office, 1884.

Horner v. Kumuliilii. 10 Haw. 174. 176 (1895).

McBryde v. Robinson. 54 Haw. 174 (1973).

Reppun v. Board of Water Supply. 65 Haw. 531 (1982).

Territory v. Gay. 31 Haw. 376, 393 (1930).

Index

Bold numerals indicate maps, and italic numerals indicate photographs.

Index

About the Author

Carol Wilcox has long been active in Hawaii's water issues. Her exposure to sugar ditch systems goes back to her early days growing up in plantation communities on Maui and Hawaii. In *Sugar Water*, she brings to life a little known but vital chapter in the history of sugar in Hawaii.